New York State Coach, Empire Edition, Mathematics, Grade 3

Triumph Learning®

New York State Coach, Empire Edition, Mathematics, Grade 3
311NY
ISBN-10: 1-60824-072-X
ISBN-13: 978-1-60824-072-2

Contributing Writer: Andie Liao
Cover Image: The Empire State Building. © Jupiterimages/Comstock

Triumph Learning® 136 Madison Avenue, 7th Floor, New York, NY 10016

Printed in the United States of America.

10 9 8 7 6 5

Table of Contents

*Standard covered at another grade level
** Grade 3 May–June Indicators

Letter to the Student

Dear Student,

Welcome to *Coach*! This book provides instruction and practice that will help you master all the important skills you need to know, and gives you practice answering the kinds of questions you will see on your state's test.

The *Coach* book is organized into chapters and lessons, and includes two Comprehensive Reviews. Before you begin the first chapter, your teacher may want you to take Comprehensive Review 1, which will help you identify skill areas that need improvement. Once you and your teacher have identified those skills, you can select the corresponding lessons and start with those. Or, you can begin with the first chapter of the book and work through to the end.

Each of the lessons has three parts. The first part walks you through the skill so you know just what it is and what it means. The second part gives you a model, or example, with hints to help your thinking about the skill. And the third part of the lesson gives you practice with the skill to see how well you understand it.

After you have finished all the lessons in the book, you can take Comprehensive Review 2 to see how much you have improved. And even if you did well on Comprehensive Review 1, you'll probably do better on Comprehensive Review 2 because practice makes perfect!

We wish you lots of success this year, and hope the *Coach* will be a part of it!

Test-Taking Checklist

Here are some tips to keep in mind when taking a test. Take a deep breath. You'll be fine!

✓ Read the directions carefully. Make sure you understand what they are asking.

✓ Do you understand the question? If not, skip it and come back to it later.

✓ Reword tricky questions. How else can the question be asked?

✓ Try to answer the question before you read the answer choices. Then pick the answer that is the most like yours.

✓ Look for words that are **bolded**, *italicized*, or <u>underlined</u>. They are important.

✓ Always look for the main idea when you read. This will help you answer the questions.

✓ Pay attention to pictures, charts, and graphs. They can give you hints.

✓ If you are allowed, use scrap paper. Take notes and make sketches if you need to.

✓ Always read all the answer choices first. Then go back and pick the best answer for the question.

✓ Be careful marking your answers. Make sure your marks are clear.

✓ Double-check your answer sheet. Did you fill in the right bubbles?

✓ Read over your answers to check for mistakes. But only change your answer if you're sure it's wrong. Your first answer is usually right.

✓ Work at your own pace. Don't go too fast, but don't go too slow either. You don't want to run out of time.

Good Luck!

New York State Math Indicators Correlation Chart

** Grade 3 May–June Indicators

Indicator	New York State Grade 3 Math Indicators	*Coach* Lesson(s)
	STRAND 1: NUMBER SENSE AND OPERATIONS	
Number Systems: Students will understand numbers, multiple ways of representing numbers, relationships among numbers, and number systems.		
3.N.1	Skip count by 25's, 50's, 100's to 1,000	1
3.N.2	Read and write whole numbers to 1,000	2
3.N.3	Compare and order numbers to 1,000	3
3.N.4	Understand the place value structure of the base ten number system: 10 ones = 1 ten 10 tens = 1 hundred 10 hundreds = 1 thousand	2
3.N.5	Use a variety of strategies to compose and decompose three-digit numbers	2
3.N.6	Use and explain the commutative property of addition and multiplication	9
3.N.7	Use 1 as the identity element for multiplication	9
3.N.8	Use the zero property of multiplication	9
3.N.9	Understand and use the associative property of addition	9
3.N.10	Develop an understanding of fractions as part of a whole unit and as parts of a collection	14
3.N.11	Use manipulatives, visual models, and illustrations to name and represent unit fractions $\left(\frac{1}{2}, \frac{1}{3}, \frac{1}{4}, \frac{1}{5}, \frac{1}{6} \text{ and } \frac{1}{10}\right)$ as part of a whole or a set of objects	13
3.N.12	Understand and recognize the meaning of numerator and denominator in the symbolic form of a fraction	13, 14
3.N.13	Recognize fractional numbers as equal parts of a whole	13, 14
**3.N.14	Explore equivalent fractions $\left(\frac{1}{2}, \frac{1}{3}, \frac{1}{4}\right)$	15
**3.N.15	Compare and order unit fractions $\left(\frac{1}{2}, \frac{1}{3}, \frac{1}{4}\right)$ and find their approximate locations on a number line	16
Number Theory		
3.N.16	Identify odd and even numbers	5
3.N.17	Develop an understanding of the properties of odd/even numbers as a result of addition or subtraction	5
Operations: Students will understand meanings of operations and procedure, and how they relate to one another.		
3.N.18	Use a variety of strategies to add and subtract 3-digit numbers (with and without regrouping)	4
3.N.19	Develop fluency with single-digit multiplication facts	8, 10
3.N.20	Use a variety of strategies to solve multiplication problems with factors up to 12 × 12	10

** Grade 3 May–June Indicators

Indicator	New York State Grade 3 Math Indicators	*Coach* Lesson(s)
3.N.21	Use the area model, tables, patterns, arrays, and doubling to provide meaning for multiplication	7, 8
3.N.22	Demonstrate fluency and apply single-digit division facts	11,12
3.N.23	Use tables, patterns, halving, and manipulatives to provide meaning for division	12
3.N.24	Develop strategies for selecting the appropriate computational and operational method in problem solving situations	17
Estimation: Students will compute accurately and make reasonable estimates.		
3.N.25	Estimate numbers up to 500	6
3.N.26	Recognize real world situations in which an estimate (rounding) is more appropriate	6
3.N.27	Check reasonableness of an answer by using estimation	6
STRAND 2: ALGEBRA		
Equations and Inequalities: Students will perform algebraic procedures accurately.		
3.A.1	Use the symbols $<, >, =$ (with and without the use of a number line) to compare whole numbers and unit fractions $\left(\frac{1}{2}, \frac{1}{3}, \frac{1}{4}, \frac{1}{5}, \frac{1}{6}, \text{ and } \frac{1}{10}\right)$	3, 18
Patterns, Relations, and Functions: Students will recognize, use, and represent algebraically patterns, relations, and functions.		
3.A.2	Describe and extend numeric $(+, -)$ and geometric patterns	19, 20
STRAND 3: GEOMETRY		
Shapes: Students will use visualization and spatial reasoning to analyze characteristics and properties of geometric shapes.		
3.G.1	Define and use correct terminology when referring to shapes (circle, triangle, square, rectangle, rhombus, trapezoid, and hexagon)	21
3.G.2	Identify congruent and similar figures	22
3.G.3	Name, describe, compare, and sort three-dimensional shapes: cube, cylinder, sphere, prism, and cone	23
3.G.4	Identify the faces on a three-dimensional shape as two-dimensional shapes	23
Transformational Geometry: Students will apply transformations and symmetry to analyze problem solving situations.		
3.G.5	Identify and construct lines of symmetry	24
STRAND 4: MEASUREMENT		
Units of Measurement: Students will determine what can be measured and how, using appropriate methods and formulas.		
3.M.1	Select tools and units (customary) appropriate for the length measured	25
3.M.2	Use a ruler/yardstick to measure to the nearest standard unit (whole and $\frac{1}{2}$ inches, whole feet, and whole yards)	25
3.M.3	Measure objects, using ounces and pounds	26
3.M.4	Recognize capacity as an attribute that can be measured	27

** Grade 3 May–June Indicators

Indicator	New York State Grade 3 Math Indicators	*Coach* Lesson(s)
3.M.5	Compare capacities (e.g., Which contains more? Which contains less?)	27
3.M.6	Measure capacity, using cups, pints, quarts, and gallons	27
Units: Students will use units to give meaning to measurements.		
3.M.7	Count and represent combined coins and dollars, using currency symbols ($0.00)	28
3.M.8	Relate unit fractions to the face of the clock: Whole = 60 minutes $\frac{1}{2}$ = 30 minutes $\frac{1}{4}$ = 15 minutes	29
Estimation: Students will develop strategies for estimating measurements.		
3.M.9	Tell time to the minute, using digital and analog clocks	29
3.M.10	Select and use standard (customary) and non-standard units to estimate measurements	25, 26, 27
STRAND 5: STATISTICS AND PROBABILITY		
Organization and Display of Data: Students will collect, organize, display, and analyze data.		
**3.S.1	Formulate questions about themselves and their surroundings	33
**3.S.2	Collect data using observation and surveys, and record appropriately	33
3.S.3	Construct a frequency table to represent a collection of data	30
3.S.4	Identify the parts of pictographs and bar graphs	31, 32
3.S.5	Display data in pictographs and bar graphs	31, 32
3.S.6	State the relationships between pictographs and bar graphs	32
Analysis of Data		
3.S.7	Read and interpret data in bar graphs and pictographs	31, 32
Predictions from Data: Students will make predictions that are based upon data analysis.		
3.S.8	Formulate conclusions and make predictions from graphs	30, 31, 32

STRAND

1 Number Sense and Operations

NYS Math Indicators

** Grade 3 May–June Indicators

1 Skip Counting

Getting the Idea

When you count forward or backward by a number other than 1, you are **skip counting.**

The number line below can help you skip count by 100s.

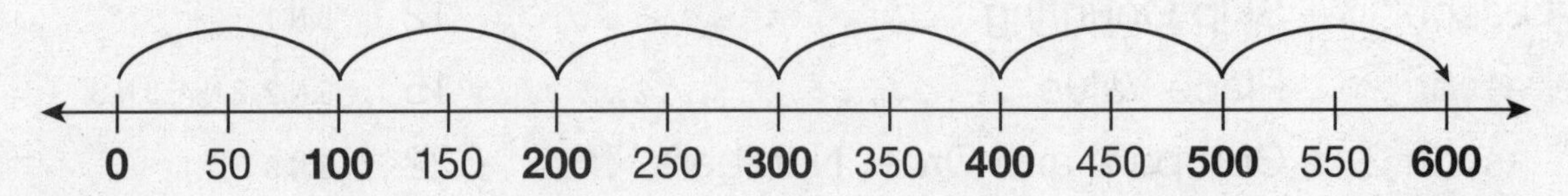

EXAMPLE 1

Each roll has 50 counters. How many counters are there in 7 rolls?

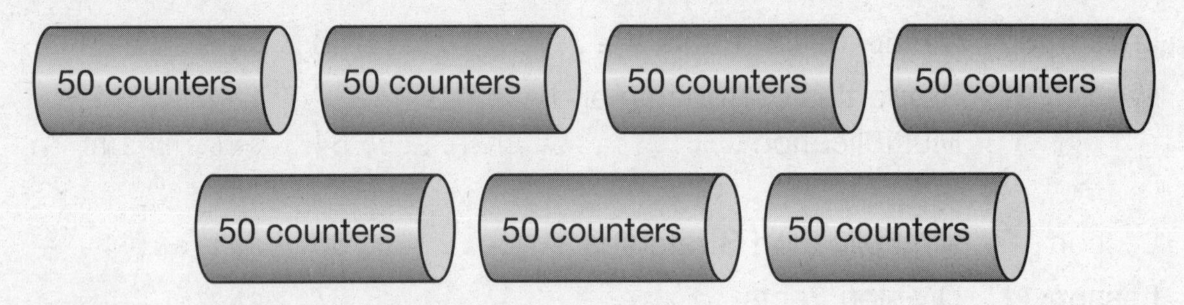

STRATEGY **Skip count by 50s.**

Start at 50 and skip count by 50s.

50, 100, 150, 200, 250, 300, 350

SOLUTION **There are 350 counters in 7 rolls.**

EXAMPLE 2

What are the next two numbers?

425, 450, 475, 500, _____, _____

STRATEGY **Use a number line.**

STEP 1 Show the numbers on the number line.

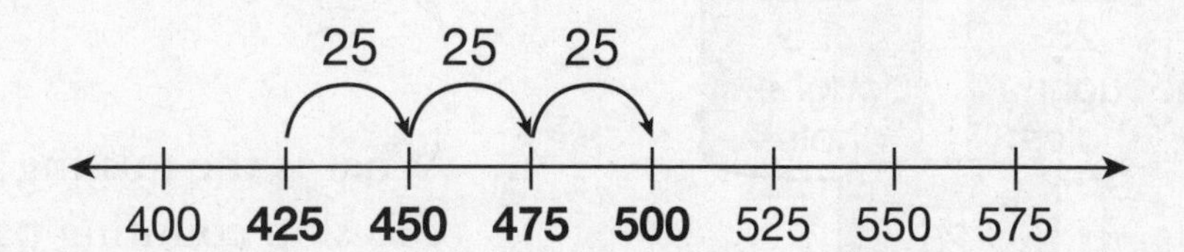

STEP 2 The numbers show skip counting by 25s.

STEP 3 Find the next two numbers. Skip count by 25s from 500.

500, 525, 550

SOLUTION **The next two numbers are 525 and 550.**

COACHED EXAMPLE

What are the missing numbers in this skip counting pattern?

280, 380, 480, 580, _____, _____, 880

THINKING IT THROUGH

The pattern is to skip count by __________.

What number comes after 580 in the pattern?

100 more than 580 is __________.

What number comes after 680 in the pattern?

100 more than 680 is__________.

The missing numbers in the pattern are __________ and __________.

Lesson Practice

Choose the correct answer.

1. Each tub has 25 doughnut holes. How many doughnut holes are in 5 tubs?

A. 525
B. 125
C. 100
D. 50

2. What number is being skip counted in the pattern below?

340, 440, 540, 640, 740, 840

A. 2
B. 25
C. 50
D. 100

3. What is the next number in this skip counting pattern?

150, 200, 250, 300, 350, ____

A. 351 **C.** 400
B. 375 **D.** 450

4. What is the missing number in this skip counting pattern?

299, 399, 499, ____, 699, 799

A. 500 **C.** 659
B. 599 **D.** 950

5. What is the next number in this skip counting pattern?

775, 800, 825, 850, 875, ____

Answer ____________

6. What is the missing number in this skip counting pattern?

250, 275, ____, 325, 350, 375

Answer ____________

2 Place Value

Getting the Idea

All numbers are made from **digits**: 0, 1, 2, 3, 4, 5, 6, 7, 8, and 9.

The number 124 has three digits. Each digit's value is based on its position in the number. This is called its **place value.**

1	**2**	**4**
↑	↑	↑
hundreds place	tens place	ones place

So 124 has 1 hundred, 2 tens, and 4 ones.

The **place value system** is based on 10s.

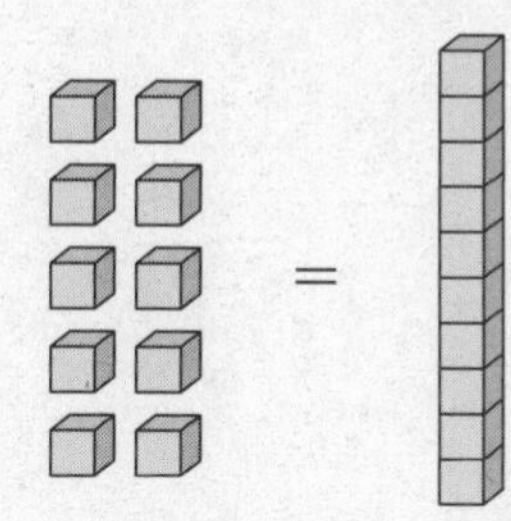

10 ones = 1 ten

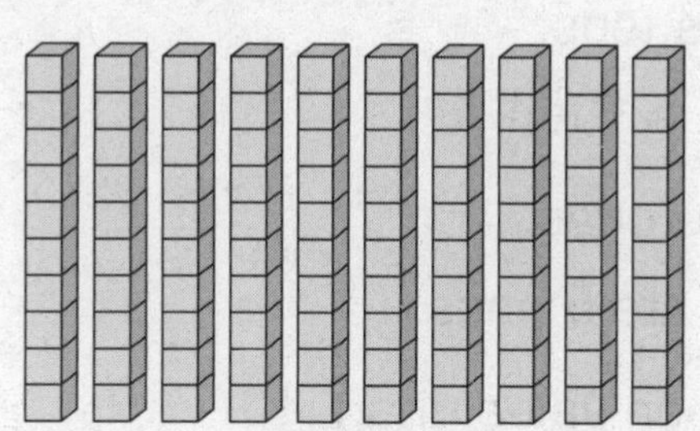

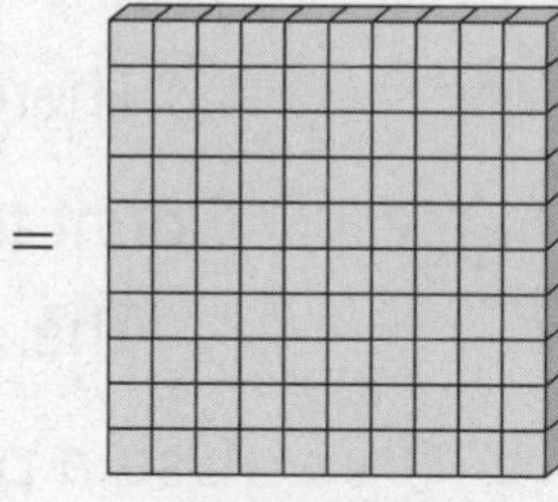

10 tens = 1 hundred

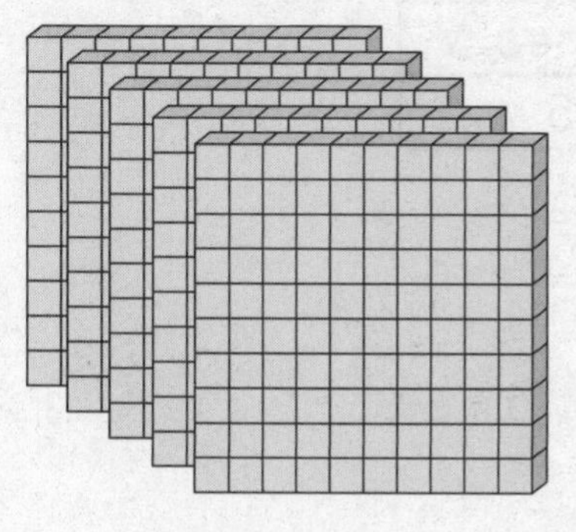

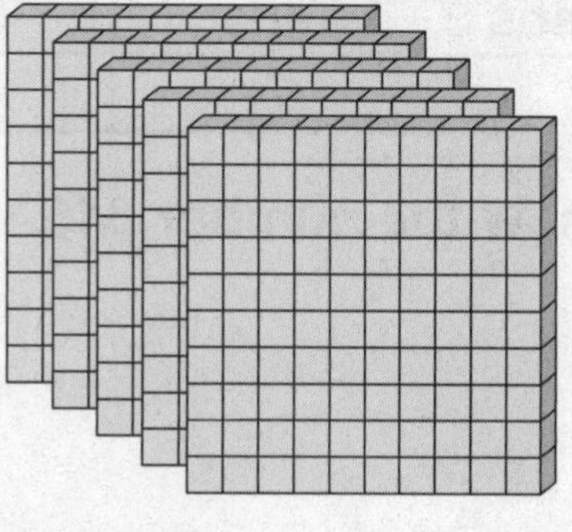

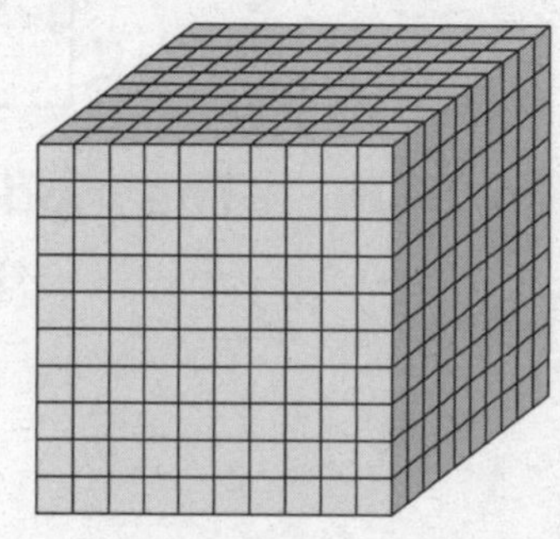

10 hundreds = 1 thousand

A **place value chart** can be used to show the value of each digit in a number.

EXAMPLE 1

What number do the models show?

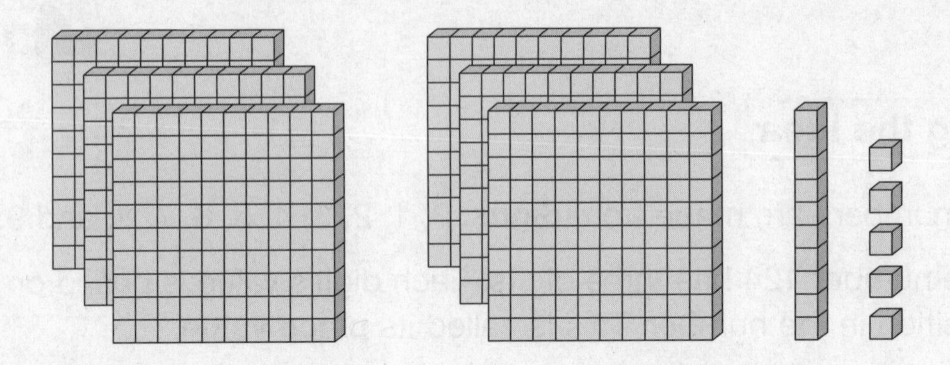

STRATEGY **Count the hundreds, tens, and ones. Use a place value chart.**

STEP 1 Count the hundreds.

There are 6 hundreds.

STEP 2 Count the tens.

There is 1 ten.

STEP 3 Count the ones.

There are 5 ones.

STEP 4 Use a place value chart.

Write each digit in the chart.

Hundreds	Tens	Ones
6	1	5

SOLUTION **The models show the number 615.**

EXAMPLE 2

What is the value of the digit 6 in 615?

STRATEGY **Look at the place value chart in Example 1.**

STEP 1 In which place is the digit 6?

The digit 6 is in the hundreds place.

STEP 2 Find the value.

6 hundreds is equal to 600.

SOLUTION **The value of the digit 6 in 615 is 600.**

A number can be written in different ways.

standard form a way to write numbers using digits: 615

word form a way to write numbers using words:
six hundred fifteen

expanded form a way to write numbers by showing the value of each digit: 600 + 10 + 5

EXAMPLE 3

What is the word form of the number 237?

STRATEGY **Use a place value chart.**

STEP 1 Write the digits in a place value chart.

Hundreds	Tens	Ones
2	3	7

STEP 2 Write the value of each digit.

2 hundreds, 3 tens, 7 ones

↓ ↓ ↓

two hundred thirty-seven

SOLUTION **The word form of 237 is two hundred thirty-seven.**

EXAMPLE 4

What is 925 in expanded form?

STRATEGY **Write the value of each digit.**

STEP 1 There are 9 hundreds.

9 hundreds = 900

STEP 2 There are 2 tens.

2 tens = 20

STEP 3 There are 5 ones.

5 ones = 5

STEP 4 Write the values of the digits.

Use a + between each value.

SOLUTION **The expanded form of 925 is 900 + 20 + 5.**

COACHED EXAMPLE

Look at the models below.

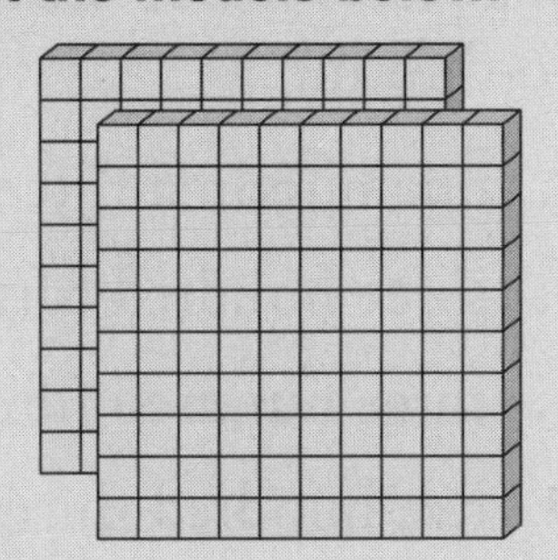
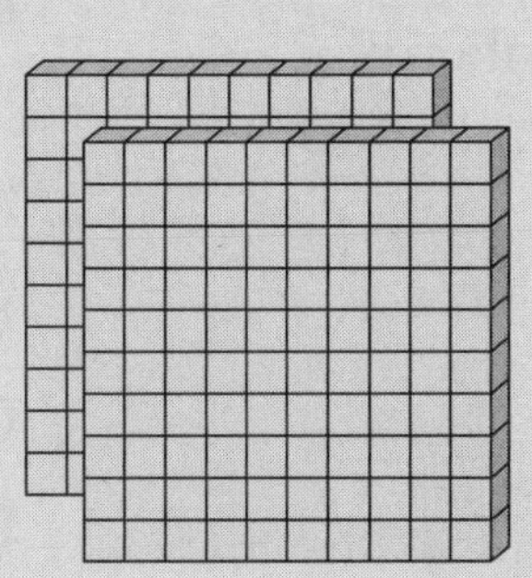
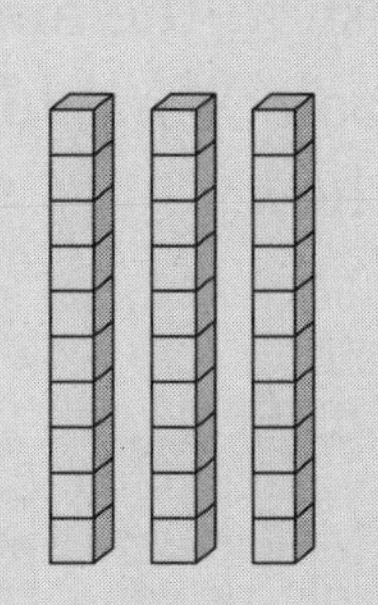
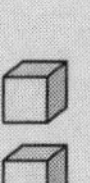

Write the number in standard form, word form, and expanded form.

THINKING IT THROUGH

Count the models.

There are _____ hundreds models.

There are _____ tens models.

There are _____ ones models.

Write the number in a place value chart.

Hundreds	Tens	Ones
4	3	2

The number in standard form is __________.

Write the hundreds part in words. ________________

Write the tens part in words. ________________

Write the ones part in words. __________

The number in word form is ______________________________________.

What is the value of the hundreds digit? __________

What is the value of the tens digit? _________

What is the value of the ones digit? __________

The number in expanded form is ____________________________________.

Lesson Practice

Choose the correct answer.

1. The models show the number of pages in a book. What number do the models show?

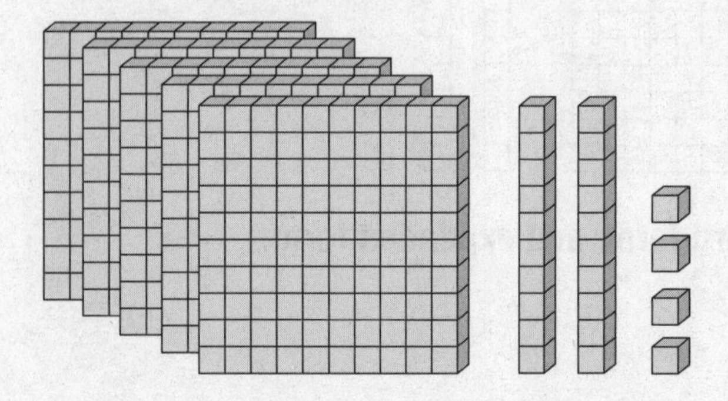

A. 124

B. 425

C. 524

D. 580

2. What number do the models show?

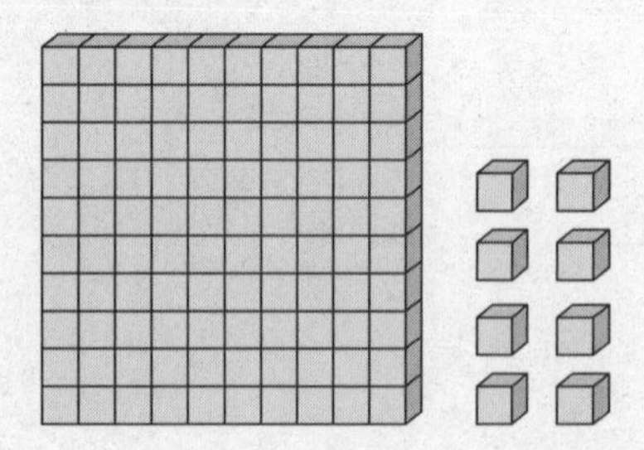

A. 18

B. 108

C. 118

D. 801

3. The Hudson River is 315 miles long. How is 315 written in word form?

A. one hundred fifteen

B. three hundred fifty

C. three hundred fifteen

D. five hundred thirteen

4. Which shows the standard form of the number nine hundred two?

A. 902

B. 912

C. 920

D. 922

5. Which shows the standard form of the number 400 + 50 + 8?

A. 854

B. 548

C. 485

D. 458

6. What is the value of the digit 7 in the number 617?

A. 7

B. 70

C. 700

D. 800

7. Write the standard form of this number.

$400 + 60 + 1$

Answer ___________

8. Write the word form of the number 790.

Answer ____________________

EXTENDED-RESPONSE QUESTION

9. David modeled the number below.

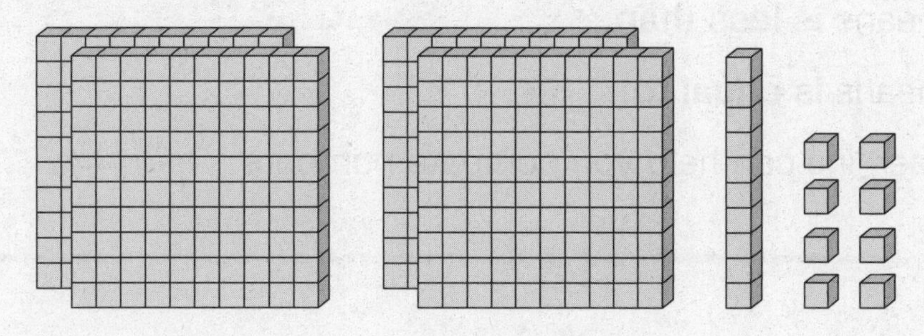

Part A Write the number in standard form.

____

Part B Write the number in word form.

____

Part C Write the number in expanded form.

____

3 Compare and Order Numbers

3.N.3, 3.A.1

Getting the Idea

Use these symbols when comparing numbers:

> means **is greater than**.

< means **is less than**.

= means **is equal to**.

A number line can help you compare numbers.

EXAMPLE 1

What symbol belongs in ◯? Write >, <, or =.

170 ◯ 150

STRATEGY **Use a number line.**

STEP 1 Place the two numbers on a number line.

STEP 2 Which number is farther to the right?

170 is to the right of 150 on the number line.

So, 170 is greater than 150.

STEP 3 Choose the correct symbol.

> means is greater than.

SOLUTION **170 (>) 150**

A place value chart can help you compare numbers. To compare, look at the places of the numbers. Start at the greatest place.

EXAMPLE 2

Kennedy Elementary School has 385 students. Jefferson Elementary School has 379 students. Which school has more students?

STRATEGY **Use a place value chart to compare the numbers.**

STEP 1 Write the numbers in a chart.

Hundreds	Tens	Ones
3	8	5
3	7	9

STEP 2 Start at the greatest place.

Compare the hundreds.

3 hundreds = 3 hundreds

Since the hundreds are equal, compare the tens.

STEP 3 Compare the tens.

8 tens is greater than 7 tens.

$385 > 379$

You do not have to compare the ones.

SOLUTION **Kennedy Elementary School has more students.**

You can also order more than two numbers.
When ordering numbers, one number is the greatest and one number is the least.

EXAMPLE 3

Tina played 3 rounds of a video game. She wrote the scores below.

842   901   849

List the scores from greatest to least.

STRATEGY **Use a place value chart.**

STEP 1 Write the numbers in a chart.

Hundreds	Tens	Ones
8	4	2
9	0	1
8	4	9

STEP 2 Start at the greatest place.

Compare the hundreds.

9 hundreds $>$ 8 hundreds

901 is the greatest number.

STEP 3 Compare 842 and 849.

8 hundreds $=$ 8 hundreds

Compare the next place.

STEP 4 Compare the tens.

4 tens $=$ 4 tens

Compare the next place.

STEP 5 Compare the ones.

2 ones $<$ 9 ones

842 is the least number.

SOLUTION **Tina's scores from greatest to least are 901, 849, 842.**

EXAMPLE 4

List the numbers below from least to greatest.

415 723 79

STRATEGY **Use place value to order the numbers.**

STEP 1 Line up the digits on the ones place.

```
415
723
 79
```

STEP 2 Compare the hundreds.

The number 79 has 0 hundreds.

79 is the least number.

STEP 3 Compare 415 and 723.

4 hundreds $<$ 7 hundreds

SOLUTION **The numbers from least to greatest are 79, 415, 723.**

COACHED EXAMPLE

What symbol belongs in ◯? Write >, <, or =.

526 ◯ 520

THINKING IT THROUGH

Use place value. Line up the numbers on the ones place.

Compare each place.

5 hundreds ◯ 5 hundreds

2 tens ◯ 2 tens

6 ones ◯ 0 ones

Is 526 greater than or less than 520? ____________________

Which symbol should you use? _____

526 ◯ 520

Lesson Practice

Choose the correct answer.

1. Which of the following is true?

A. $366 < 309$

B. $980 > 798$

C. $741 = 147$

D. $563 < 535$

2. Which lists the numbers from greatest to least?

A. 435 543 584

B. 543 354 435

C. 354 543 435

D. 543 435 354

3. Which number is between 882 and 886?

A. 883

B. 882

C. 881

D. 828

4. Which number makes this sentence true?

$675 < 6\square4$

A. 3

B. 5

C. 7

D. 8

5. The table shows the number of CDs sold at a store in a week.

CD Sales

Type	Number of CDs
Rock	372
R&B	355
Classical	93
Pop	407

Which type of CD sold the most?

A. rock

B. R&B

C. classical

D. pop

6. Which symbol belongs in the ◯?

Write $>$, $<$, or $=$.

793 ◯ 806

Answer ______________

7. Order the numbers from least to greatest.

374 805 23 986

Answer ______________

Lesson

Add and Subtract Whole Numbers

Getting the Idea

When you **add** numbers, you combine quantities.

Here are the parts to an addition sentence.

$$\begin{array}{rcl} 247 & \leftarrow & \textbf{addend} \\ +\,351 & \leftarrow & \textbf{addend} \\ \hline 598 & \leftarrow & \textbf{sum} \end{array}$$

Add the digits from right to left. If the sum of a column is 10 or greater, you will have to **regroup** 10 of one unit to 1 of the next greatest unit.

EXAMPLE 1

Sheri sold 362 raffle tickets. Kevin sold 394 tickets.
How many tickets did they sell in all?

STRATEGY **Add from right to left.**

STEP 1 Line up the digits on the ones place.

$$\begin{array}{r} 362 \\ +\,394 \\ \hline \end{array}$$

STEP 2 Add the ones.

$2 + 4 = 6$

$$\begin{array}{r} 36\mathbf{2} \\ +\,39\mathbf{4} \\ \hline \mathbf{6} \end{array}$$

STEP 3 Add the tens. Regroup.

$6 + 9 = 15$

15 tens = 1 hundred 5 tens

$$\begin{array}{r} \scriptstyle 1 \\ 3\mathbf{6}2 \\ +\,3\mathbf{9}4 \\ \hline \mathbf{5}6 \end{array}$$

STEP 4 Add the hundreds.

$1 + 3 + 3 = 7$

$$\begin{array}{r} \scriptstyle 1 \\ \mathbf{3}62 \\ +\,\mathbf{3}94 \\ \hline \mathbf{7}56 \end{array}$$

SOLUTION **Sherri and Kevin sold 756 tickets in all.**

When you **subtract**, you take away an amount.

Here are the parts to a subtraction sentence.

$$\begin{array}{r c l} 5\,4\,8 & \rightarrow & \textbf{minuend} \\ -\,3\,1\,5 & \rightarrow & \textbf{subtrahend} \\ \hline 2\,3\,3 & \rightarrow & \textbf{difference} \end{array}$$

Subtract the digits from right to left. Sometimes you may need to regroup.

EXAMPLE 2

827 − 368 = □

STRATEGY **Subtract from right to left.**

STEP 1 Line up the digits on the ones place.

There are not enough ones to subtract.

Regroup 1 ten as 10 ones.

$$\begin{array}{r} {}^{1\,17} \\ 8\,\cancel{2}\,\cancel{7} \\ -\,3\,6\,8 \\ \hline \end{array}$$

STEP 2 Subtract the ones.

17 − 8 = 9

$$\begin{array}{r} {}^{1\,17} \\ 8\,\cancel{2}\,\cancel{7} \\ -\,3\,6\,\mathbf{8} \\ \hline \mathbf{9} \end{array}$$

STEP 3 There are not enough tens.

Regroup 1 hundred as 10 tens.

Subtract the tens.

11 − 6 = 5

$$\begin{array}{r} {}^{11} \\ {}^{7\,\cancel{1}\,17} \\ \cancel{8}\,\cancel{2}\,\cancel{7} \\ -\,3\,\mathbf{6}\,8 \\ \hline \mathbf{5}\,9 \end{array}$$

STEP 4 Subtract the hundreds.

7 − 3 = 4

$$\begin{array}{r} {}^{11} \\ {}^{7\,\cancel{1}\,17} \\ \cancel{8}\,\cancel{2}\,\cancel{7} \\ -\,\mathbf{3}\,6\,8 \\ \hline \mathbf{4}\,5\,9 \end{array}$$

SOLUTION **827 − 368 = 459**

Sometimes when you subtract, you may need to regroup across zeros.

EXAMPLE 3

A theater has 600 seats. People are sitting in 557 of the seats. The rest of the seats are empty. How many seats are empty?

STRATEGY **Subtract from right to left.**

STEP 1 There are not enough ones or tens to subtract.

Regroup 1 hundred as 10 tens. Then regroup 1 ten as 10 ones.

$$\begin{array}{r} & & \mathbf{9} & \\ & \mathbf{5} & \mathbf{\not{10}} & \mathbf{10} \\ & \not{6} & \not{0} & \not{0} \\ - & 5 & 5 & 7 \\ \hline \end{array}$$

STEP 2 Subtract the ones.

$10 - 7 = 3$

$$\begin{array}{r} & & 9 & \\ & 5 & \not{10} & \mathbf{10} \\ & \not{6} & \not{0} & \not{0} \\ - & 5 & 5 & \mathbf{7} \\ \hline & & & \mathbf{3} \end{array}$$

STEP 3 Subtract the tens.

$9 - 5 = 4$

$$\begin{array}{r} & & \mathbf{9} & \\ & 5 & \not{10} & 10 \\ & \not{6} & \not{0} & \not{0} \\ - & 5 & \mathbf{5} & 7 \\ \hline & & \mathbf{4} & 3 \end{array}$$

STEP 4 Subtract the hundreds.

$5 - 5 = 0$

$$\begin{array}{r} & & 9 & \\ & \mathbf{5} & \not{10} & 10 \\ & \not{6} & \not{0} & \not{0} \\ - & \mathbf{5} & 5 & 7 \\ \hline & & 4 & 3 \end{array}$$

SOLUTION **There are 43 empty seats.**

You can use addition to check the difference.

$$\begin{array}{r} 600 \\ -\,557 \\ \hline 43 \end{array} \qquad \begin{array}{r} 557 \\ +\;\;43 \\ \hline 600 \end{array}$$

COACHED EXAMPLE

Eric's school has 183 boys and 194 girls as students. There are also 86 adults working at the school. How many more students than adults are in Eric's school?

THINKING IT THROUGH

First, find how many students there are in all.

Add from right to left.

$$\begin{array}{r} 183 \\ +\,194 \\ \hline \end{array}$$

Add the ones. _____ + _____ = _____ ones

Add the tens. _____ + _____ = _____ tens

Regroup _____ tens as ____ hundred _____ tens.

Add the hundreds. _____ + _____ + _____ = _____

Find how many more students than adults.

Subtract from right to left.

Subtract the ones. _____ − _____ = _____ one

There are not enough tens to subtract.

Regroup _____ hundred and 7 tens as _____ tens.

Subtract the tens. _____ − _____ = _____ tens

Subtract the hundreds. _____ − _____ = _____ hundreds

There are ___________ more students than adults at Eric's school.

Lesson Practice

Choose the correct answer.

1.

$$\begin{array}{r} 484 \\ +\,275 \\ \hline \end{array}$$

A. 759

B. 719

C. 659

D. 209

2.

$$\begin{array}{r} 539 \\ -\,246 \\ \hline \end{array}$$

A. 213

B. 293

C. 373

D. 785

3.

$$\begin{array}{r} 703 \\ -\,358 \\ \hline \end{array}$$

A. 345

B. 355

C. 445

D. 455

4. New York City has an area of 304 square miles. Philadelphia has an area of 127 square miles. What is the total area of the two cities?

A. 421 square miles

B. 431 square miles

C. 433 square miles

D. 531 square miles

5. A movie theater sold 692 tickets to the evening show. It sold 385 tickets to the afternoon show. How many more tickets were sold for the evening show than the afternoon show?

A. 207

B. 217

C. 307

D. 317

6. Which digit goes in the □ to make the sentence true?

$$\begin{array}{r} 3\ 7\ 6 \\ +\,1\ \square\ 2 \\ \hline 5\ 0\ 8 \end{array}$$

A. 0

B. 3

C. 4

D. 5

7. Texas has 254 counties. New York has 192 counties fewer than Texas. How many counties does New York have?

Answer ______

8. The school band played two concerts. The first concert had 358 attendees. The second concert had 475 attendees. How many attendees were at the two concerts in all?

Answer ______

EXTENDED-RESPONSE QUESTION

9. The table shows the number of trading cards Richie has.

Number of Trading Cards

Type	Number of Cards
Basketball	173
Football	126
Baseball	429

Part A How many basketball cards and football cards does Richie have in all?

Part B How many more baseball cards does Richie have than basketball cards and football cards combined?

Part C Explain how you solved each part of the problem.

______

______

______

Lesson

5 Even and Odd Numbers

3.N.16, 3.N.17

Getting the Idea

An **even number** is a number that has a 0, 2, 4, 6, or 8 in the ones place. An even number can be separated into two equal groups.

An **odd number** is a number that has a 1, 3, 5, 7 or 9 in the ones place. An odd number cannot be separated into two equal groups.

EXAMPLE 1

Is 35 an even number or an odd number?

STRATEGY **Look at the digit in the ones place.**

The digit in the ones place is 5.

The number is odd.

SOLUTION **35 is an odd number.**

EXAMPLE 2

Is 180 an even number or an odd number?

STRATEGY **Look at the digit in the ones place.**

The digit in the ones place is 0.

The number is even.

SOLUTION **180 is an even number.**

You can tell whether a sum or difference of two numbers will be odd or even. Use the tables below.

Addition

addend	+	addend	=	sum
odd	+	odd	=	even
odd	+	even	=	odd
even	+	odd	=	odd
even	+	even	=	even

Subtraction

minuend	–	subtrahend	=	difference
odd	–	odd	=	even
odd	–	even	=	odd
even	–	odd	=	odd
even	–	even	=	even

EXAMPLE 3

Is the difference of 34 – 11 an even or odd number?

STRATEGY **Decide if each number is odd or even. Use the table.**

STEP 1 Identify 34 and 11 as odd or even.

34 has a 4 in the ones place. It is even.

11 has a 1 in the ones place. It is odd.

STEP 2 Use the table.

even – odd = odd

STEP 3 Check if the difference is odd.

34 – 11 = 23

23 has a 3 in the ones place.

It is an odd number.

SOLUTION **The difference of 34 – 11 is an odd number.**

COACHED EXAMPLE

On game day, there were 85 second-grade students and 67 third-grade students. A game was played where each student needed a partner. Did each student have a partner for the game?

THINKING IT THROUGH

Decide if each addend is odd or even.

85 has a(n) _____ in the ones place.

85 is an __________ number.

67 has a(n) _____ in the ones place.

67 is an __________ number.

__________ number + __________ number = __________ number

If the sum of 85 + 67 is an __________ number, then each student had a partner.

Since the sum is an __________ number, each student __________ a partner.

Lesson Practice

Choose the correct answer.

1. Which is an even number?

   A. 11
   B. 125
   C. 354
   D. 659

2. Which is an odd number?

   A. 130
   B. 258
   C. 592
   D. 727

3. Which shows all even numbers?

   A. 120 248 374 402
   B. 114 246 329 416
   C. 290 330 342 473
   D. 182 265 384 410

4. Which shows all odd numbers?

   A. 377 285 797 800
   B. 171 383 399 505
   C. 280 291 403 615
   D. 467 580 693 809

5. I am an even number with an odd number of digits. Which number am I?

   A. 20
   B. 315
   C. 498
   D. 509

6. I am an odd number. My hundreds and tens digits are even. Which number am I?

   A. 935
   B. 810
   C. 624
   D. 463

7. Is the sum of 240 + 545 odd or even?

   Answer ____________

8. Is the difference of 543 – 299 odd or even?

   Answer ____________

Lesson

6 Estimation

3.N.25, 3.N.26, 3.N.27

Getting the Idea

Sometimes you do not need to find an exact answer to a problem. If the problem asks for *about* how many or *about* how much, you can use an **estimate**. An estimate is a number close to the exact answer. One way to estimate is to **round** the numbers to the nearest ten or hundred.

Remember these **rounding rules**:
Look at the digit to the right of the place you are rounding to.

- If the digit is less than 5, round down.
- If the digit is 5 or greater, round up.

EXAMPLE 1

Polk Elementary School has 413 students. Grant Elementary School has 278 students. About how many more students attend Polk than Grant?

STRATEGY **Decide if you need an exact answer or an estimate. Then solve the problem.**

STEP 1 Decide if you need an exact answer or an estimate.

The problem asks, "*about* how many," so estimate.

STEP 2 Round the numbers to the nearest hundred.

Look at the digits in the tens place.

$1 < 5$, so 4**1**3 rounds down to 400.

$7 > 5$, so 2**7**8 rounds up to 300.

STEP 3 The problem asks, "about how many *more*," so subtract.

$400 - 300 = 100$

SOLUTION **About 100 more students attend Polk Elementary School than Grant Elementary School.**

EXAMPLE 2

Three tour buses are on their way to the stadium. The first bus has 43 people. The second bus has 47 people. The third bus has 38 people. Estimate how many people are on the buses in all.

STRATEGY **Round to the nearest ten. Then add.**

STEP 1 Round each number to the nearest ten.

Look at the digit in the ones place.

43 rounds down to 40.

47 rounds up to 50.

38 rounds up to 40.

STEP 2 Add the rounded numbers.

$40 + 50 + 40 = 130$

SOLUTION **There are about 130 people on the buses in all.**

You can use estimation to check if an answer is reasonable. This means that you are checking to see if an answer makes sense.

EXAMPLE 3

Molly had $87 in her wallet. She bought a pair of pants for $32. She estimated that she had about $30 left. Is Molly's estimate reasonable?

STRATEGY **Round each number to the greatest place. Then subtract.**

STEP 1 Round each number to the nearest ten dollars.

$87 rounds up to $90 because $7 > 5$.

$32 rounds down to $30 because $2 < 5$.

STEP 2 Subtract the rounded numbers.

$\$90 - \$30 = \$60$

$60 is not close to $30.

SOLUTION **Molly's estimate is not reasonable.**

COACHED EXAMPLE

This week, 128 people visited the skate park. Last week, 176 people visited the park. About how many people in all visited the park in two weeks?

THINKING IT THROUGH

"About how many" tell you to find an estimate.

Use the rounding rules to round each number to the nearest hundred.

Round 128 to the nearest hundred.

$2 < 5$, so round 128 up or down? ____________

To the nearest hundred, 128 rounds to __________.

Round 176 to the nearest hundred.

$7 > 5$, so round 176 up or down? ____________

To the nearest hundred, 176 rounds to __________.

________________ the rounded numbers to find how many people in all.

__________ + __________ = __________

About __________ people in all visited the park in two weeks.

Lesson Practice

Choose the correct answer.

1. John is reading a book with 332 pages in all. He has read 128 pages. About how many more pages are left for John to read to finish the book?

A. 100

B. 200

C. 300

D. 400

2. A amusement park had 376 visitors in the morning. The park had another 145 visitors in the evening. About how many visitors did the park have that day?

A. 200

B. 300

C. 400

D. 500

3. Estimate: 184 + 245

A. 300

B. 400

C. 500

D. 600

4. Rose has 218 coins in her collection. Michelle has 181 coins in her collection. About how many coins do they have in all?

A. 200

B. 300

C. 400

D. 500

5. Estimate: 404 − 159

A. 50

B. 100

C. 200

D. 400

6. Felicia had $214 in her savings account. Then she put $88 into the account. Which is the best estimate of the amount of money in the savings account now?

A. $200

B. $300

C. $400

D. $500

7. A farmer planted 83 tomato seeds and 76 pumpkin seeds. To the nearest ten, about how many tomato and pumpkin seeds did the farmer plant?

A. 120

B. 160

C. 200

D. 400

8. What is the estimated sum?

182 + 115

Answer ___________

9. Ms. Jenkins spent $52 on a chair and $148 on a table. About how much more does the table cost than the chair?

Answer ___________

EXTENDED-RESPONSE QUESTION

10. Each month, Mr. Adams pays $495 in rent and $279 for his car payment.

Part A About how much more is his rent than his car payment?

____

Part B On the lines below, explain why your estimate in Part A is reasonable.

____

____

____

Lesson

7 Understand Multiplication

Getting the Idea

Multiplication is a form of repeated addition. When you **multiply**, you join equal groups.

Here are the parts of a multiplication sentence.

3	×	2	=	6
factor		**factor**		**product**

An **array** shows equal groups of objects in rows and columns.

EXAMPLE 1

Write a multiplication sentence for this array.

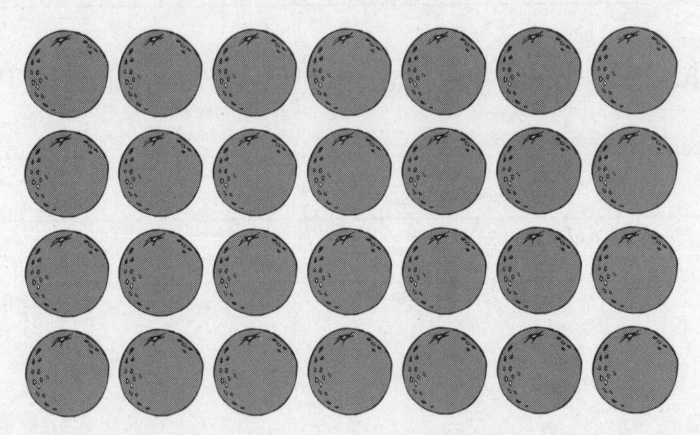

STRATEGY **Count the number of rows and the number in each row.**

STEP 1 Count the number of rows. Count the number of oranges in each row.

There are 4 rows.

Each row has 7 oranges.

STEP 2 Find the total number of oranges.

4 groups of 7 total 28

STEP 3 Write the multiplication fact.

4 rows of 7 total 28

$4 \times 7 = 28$

SOLUTION **The array shows the multiplication sentence $4 \times 7 = 28$.**

You can also use repeated addition to solve multiplication.

EXAMPLE 2

How many pencils are there in all? Find the total and write the multiplication sentence.

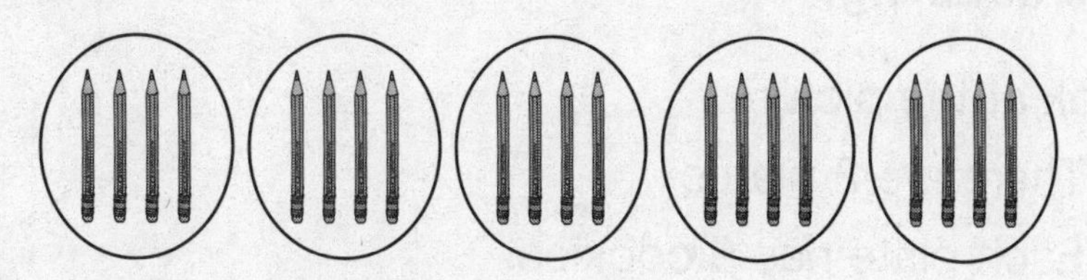

STRATEGY **Use repeated addition.**

STEP 1 Count the number of pencils in each group and the number of groups.

There are 4 pencils in each group.

There are 5 groups.

STEP 2 Add 4 five times to find the total number of pencils.

$4 + 4 + 4 + 4 + 4 = 20$

STEP 3 Write the multiplication sentence.

5 groups of 4 total 20

$5 \times 4 = 20$

SOLUTION **There are 20 pencils in all.**
$5 \times 4 = 20$

When you multiply by 2, use doubling to find the product.

EXAMPLE 3

How many cookies are there in all?

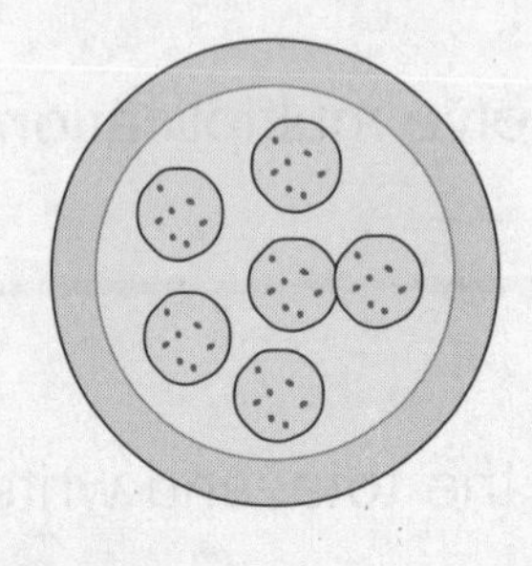
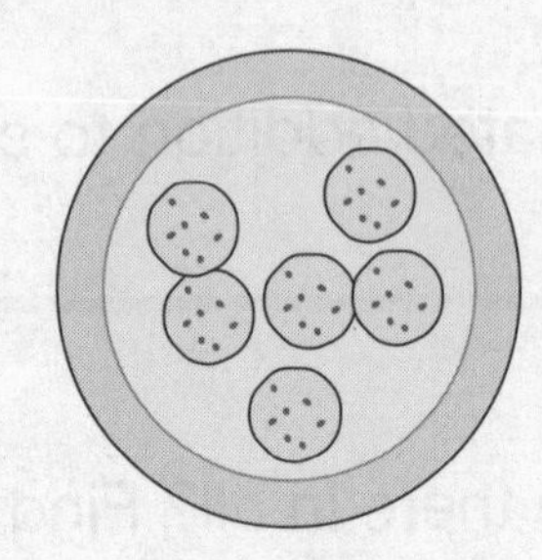

STRATEGY **Use doubling.**

STEP 1 Look at the picture.

There are 2 plates.

Each plate has 6 cookies.

STEP 2 Double 6.

$6 + 6 = 12$.

STEP 3 Write a multiplication sentence.

2 groups of 6 total 12.

$2 \times 6 = 12$

SOLUTION **There are 12 cookies in all.**
$\mathbf{2 \times 6 = 12}$

EXAMPLE 4

Carlos bought 3 boxes of golf balls. Each box has 6 balls. How many golf balls did Carlos buy?

STRATEGY **Use skip counting.**

STEP 1 Show the problem as multiplication.

3 boxes of 6 balls = 3 groups of 6

$3 \times 6 = ?$

STEP 2 Skip count by 6 three times.

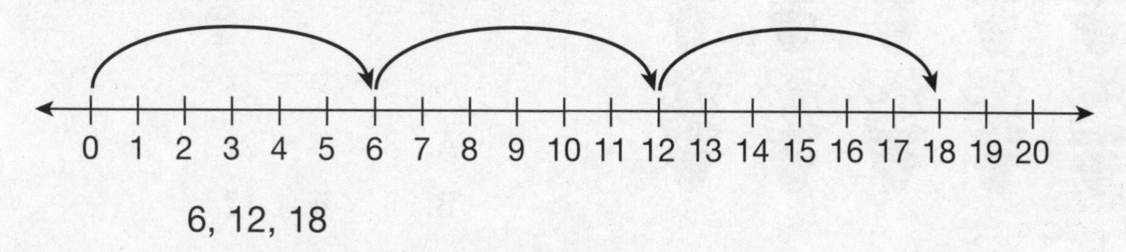

6, 12, 18

SOLUTION **Carlos bought 18 golf balls. $3 \times 6 = 18$**

COACHED EXAMPLE

Write a multiplication sentence for this model.

THINKING IT THROUGH

Count the number of rows. _____

Count the number of squares in each row. _____

Use skip counting to find the total number of squares. _____ _____ _____

The model shows the multiplication sentence ____________________.

Lesson Practice

Choose the correct answer.

1. Which multiplication sentence does this array show?

A. $3 \times 2 = 6$
B. $3 \times 5 = 15$
C. $5 \times 2 = 10$
D. $5 \times 5 = 25$

2. Which multiplication sentence does this model show?

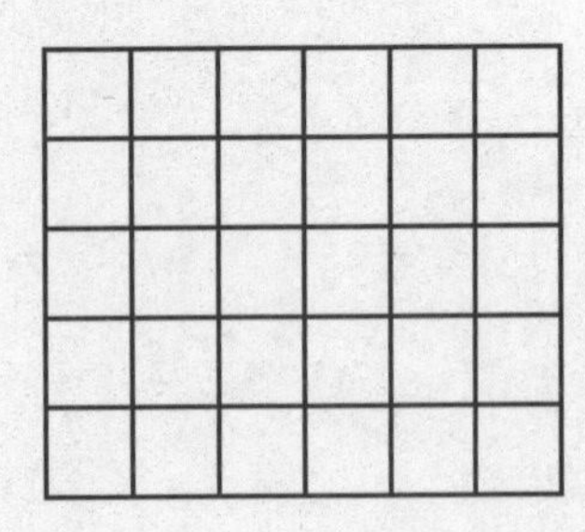

A. $5 \times 5 = 25$
B. $5 \times 1 = 5$
C. $6 \times 3 = 18$
D. $5 \times 6 = 30$

3. Which multiplication sentence shows the total number of children?

A. $3 \times 7 = 21$
B. $3 \times 4 = 12$
C. $7 \times 1 = 7$
D. $4 \times 6 = 24$

4. Which multiplication sentence does this model show?

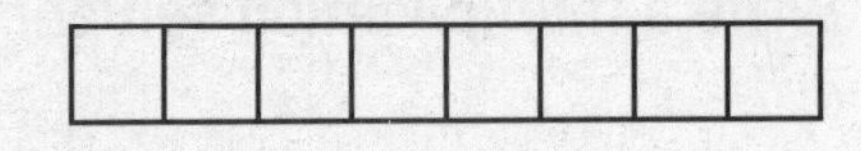

A. $2 \times 4 = 8$
B. $1 \times 8 = 8$
C. $8 \times 8 = 64$
D. $7 \times 2 = 14$

5. Which multiplication sentence does this array show?

○ ○ ○ ○ ○

○ ○ ○ ○ ○

A. $2 \times 10 = 20$

B. $2 \times 5 = 10$

C. $5 \times 5 = 25$

D. $2 \times 6 = 12$

6. Which multiplication sentence does the repeated addition below show?

$2 + 2 + 2 + 2$

A. $2 \times 8 = 16$

B. $2 \times 6 = 12$

C. $2 \times 4 = 8$

D. $2 \times 2 = 4$

7. Which repeated addition can be used to solve 4×6?

A. $4 + 6 + 4 + 6$

B. $4 + 4 + 4 + 4$

C. $6 + 6 + 6 + 6$

D. $2 + 4 + 6 + 8$

8. Anna skip counts by 2s eight times to solve a multiplication sentence.

Which multiplication sentence is she solving?

A. $2 \times 2 = 4$

B. $8 \times 8 = 64$

C. $2 \times 8 = 16$

D. $6 \times 2 = 12$

9. Write a multiplication sentence for this array.

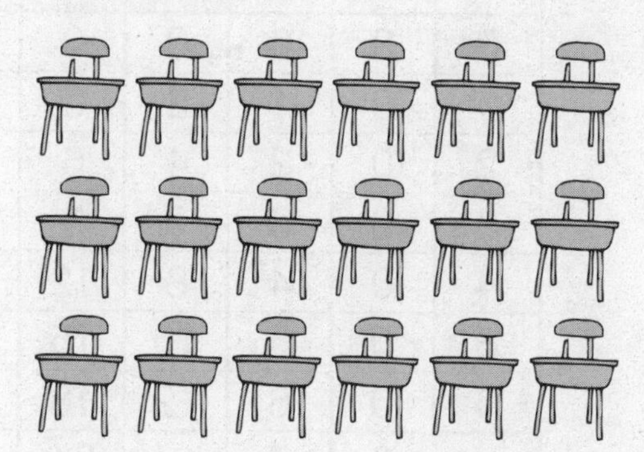

Answer ______________________

10. Write a multiplication sentence for this repeated addition.

$5 + 5 + 5 + 5 + 5 + 5 + 5 + 5$

Answer ______________________

Lesson

8 Multiplication Facts

3.N.19, 3.N.21

Getting the Idea

You can use a multiplication table to memorize basic facts.
The factors are in the top row and the first column.
To find a product, look for the row of one factor and the column of the other factor. The number in the box where they meet is the product.

×	0	1	2	3	4	5	6	7	8	9	10
0	0	0	0	0	0	0	0	0	0	0	0
1	0	1	2	3	4	5	6	7	8	9	10
2	0	2	4	6	8	10	12	14	16	18	20
3	0	3	6	9	12	15	18	21	24	27	30
4	0	4	8	12	16	20	24	28	32	36	40
5	0	5	10	15	20	25	30	35	40	45	50
6	0	6	12	18	24	30	36	42	48	54	60
7	0	7	14	21	28	35	42	49	56	63	70
8	0	8	16	24	32	40	48	56	64	72	80
9	0	9	18	27	36	45	54	63	72	81	90
10	0	10	20	30	40	50	60	70	80	90	100

EXAMPLE 1

$5 \times 10 = \square$

STRATEGY **Use a multiplication table.**

STEP 1 Find the 5s row. Find the 10s column.

STEP 2 Find the box where the row and column meet.

The product, 50, is inside that box.

SOLUTION $\mathbf{5 \times 10 = 50}$

EXAMPLE 2

$4 \times 3 = \square$

STRATEGY **Use a multiplication table.**

STEP 1 Find the 4s row. Find the 3s column.

STEP 2 Find the box where the row and the column meet.

The product, 12, is inside that box.

SOLUTION $\mathbf{4 \times 3 = 12}$

EXAMPLE 3

Mary has 3 rows of marbles. Each row has 5 marbles. How many marbles does she have in all?

STRATEGY **Find the factors in the problem. Use a multiplication table.**

STEP 1 Identify the factors.

There are 3 rows.

There are 5 marbles in each row.

The factors are 3 and 5.

STEP 2 Use the multiplication table.

Find the box where the 3s row and the 5s column meet.

The product, 15, is inside that box.

SOLUTION **Mary has 15 marbles in all.**

EXAMPLE 4

Mr. Cole bought 5 T-shirts. Each T-shirt cost $5. How much did the 5 T-shirts cost in all?

STRATEGY **Use repeated addition.**

STEP 1 Show the problem as multiplication.

5 shirts at $5 each = 5 groups of 5

$5 \times \$5 = ?$

STEP 2 Add 5 five times.

$5 + 5 + 5 + 5 + 5 = 25$

SOLUTION **The 5 T-shirts cost $25 in all.**

COACHED EXAMPLE

Celine brushes her teeth 3 times each day.
How many times does she brush her teeth in a week?
(There are 7 days in a week.)

THINKING IT THROUGH

Write this problem as _______ × _______ = ?

Use repeated addition. ____

What is the total? _____

Celine brushes her teeth _______ times a week.

Lesson Practice

Choose the correct answer.

1. $3 \times 3 = \square$

A. 6
B. 9
C. 11
D. 30

2. $5 \times 2 = \square$

A. 3
B. 7
C. 10
D. 15

3. $2 \times 9 = \square$

A. 11
B. 18
C. 29
D. 92

4. $5 \times 1 = \square$

A. 5
B. 6
C. 10
D. 15

5. $3 \times 2 = \square$

A. 5
B. 6
C. 23
D. 32

6. $4 \times 8 = \square$

A. 12
B. 24
C. 32
D. 48

7. $1 \times 9 = \square$

A. 0
B. 1
C. 9
D. 10

8. $2 \times 8 = \square$

A. 10
B. 16
C. 28
D. 60

9. $4 \times 4 = \square$

A. 4

B. 8

C. 12

D. 16

10. $3 \times 9 = \square$

A. 39

B. 27

C. 16

D. 12

11. Which multiplication sentence does **not** have the same product as the others?

A. $5 \times 3 = \square$

B. $2 \times 6 = \square$

C. $4 \times 3 = \square$

D. $3 \times 4 = \square$

12. Five parents are driving the students to the play. Each parent takes 4 students. How many students are going to the play?

A. 9

B. 15

C. 20

D. 25

13. Juan baked 5 trays of muffins. Each tray holds 6 muffins. How many muffins did Juan bake in all?

A. 11

B. 25

C. 30

D. 50

14. The table shows the cost of grapes.

Grape Prices

Weight	Total Cost
1 pound	$2
2 pounds	$4
3 pounds	$6
4 pounds	

Each pound of grapes costs $2. Which multiplication sentence shows how much 4 pounds of grapes cost?

A. $\$4 \times 4 = \$16$

B. $\$2 \times 4 = \$8$

C. $\$7 \times 1 = \$7$

D. $\$2 \times 2 = \$4$

15. What is the product of 5×9?

Answer __________

16. Rachel used 3 slices of cheese in each sandwich. She made 8 sandwiches. How many slices of cheese did she use in all?

Answer __________

EXTENDED-RESPONSE QUESTION

17. Candice has a shelf with 4 rows.
Each row has 5 books.

Part A Draw an array to show the problem.

Part B What is the total number of books on the shelf?
Write a multiplication sentence.

____

Part C Explain how you solved each part of the problem.

____

____

____

Lesson

Properties of Addition and Multiplication

3.N.6, 3.N.7, 3.N.8, 3.N.9

Getting the Idea

Using addition properties can make it easier for you to add numbers.

The **commutative property of addition** says that changing the order of the addends does not change the sum.

$$2 + 4 = 4 + 2$$

$$6 = 6$$

EXAMPLE 1

Which number makes this number sentence true?

$12 + 18 = 18 + \square$

STRATEGY **Use the commutative property of addition.**

The commutative property of addition says that changing the order of the addends does not change the sum.

$12 + 18 = 30$, so $18 + 12 = 30$

SOLUTION **The number 12 makes this sentence true.**

The **associative property of addition** says that changing the grouping of the addends does not change the sum.

$$(4 + 3) + 5 = 4 + (3 + 5)$$

$$7 + 5 = 4 + 8$$

$$12 = 12$$

EXAMPLE 2

Add.

$3 + (17 + 6) = \square$

STRATEGY **Use the associative property of addition.**

STEP 1 Change the grouping of the addends.

Use mental math. Think: $3 + 17 = 20$

$3 + (17 + 6) = (3 + 17) + 6$

STEP 2 Find the sum.

$(3 + 17) + 6$

$20 \quad + 6 = 26$

SOLUTION **$3 + (17 + 6) = 26$**

The **commutative property of multiplication** says that changing the order of the factors does not change the product.

$3 \times 4 = 12$ $\quad$ $4 \times 3 = 12$

EXAMPLE 3

Which number makes this number sentence true?

$2 \times 5 = \square \times 2$

STRATEGY **Use the commutative property of multiplication.**

Changing the order of the factors does not change the product.

$2 \times 5 = 10$

$5 \times 2 = 10$

The products are the same.

SOLUTION **The number 5 makes the sentence true.**

The **zero property of multiplication** says that the product of any number and 0 is 0.

For example, $5 \times 0 = 0$.

EXAMPLE 4

Multiply.

$4 \times 0 = \square$

STRATEGY **Use the zero property of multiplication.**

STEP 1 Look at the factors.

One factor is 0.

STEP 2 Use the zero property of multiplication.

When a factor is multiplied by 0, the product is 0.

$4 \times 0 = 0$

SOLUTION $\mathbf{4 \times 0 = 0}$

The **identity property of multiplication** says that the product of any number and 1 is that same number.

For example, $10 \times 1 = 10$.

EXAMPLE 5

What is the missing number?

$3 \times \square = 3$

STRATEGY **Use the identity property of multiplication.**

STEP 1 Look at the numbers in the sentence.

One factor is 3. The product is 3.

STEP 2 Use the identity property of multiplication.

When a factor is multiplied by 1, the product is that factor.

$3 \times 1 = 3$

SOLUTION **The missing number is 1.**

COACHED EXAMPLE

If $4 \times 9 = 36$, what is the missing factor in the number sentence below?

$\square \times 4 = 36$

THINKING IT THROUGH

What is the product in the number sentence $4 \times 9 = 36$? ____________

What is the product in the number sentence $\square \times 4 = 36$? ____________

Are the products the same? ____________

What are the two factors in the number sentence $4 \times 9 = 36$? _____ and ______

What property of multiplication says that multiplying the factors in a different order does not change the product? ________________ ________________

What is the missing factor in the number sentence $\square \times 4 = 36$? _____

The missing factor in $\square \times 4 = 36$ is ____________.

Lesson Practice

Choose the correct answer.

1. Which number makes this number sentence true?

$15 + 16 = \square + 15$

A. 31
B. 21
C. 16
D. 15

2. Which number makes this number sentence true?

$4 \times \square = 4$

A. 1
B. 4
C. 5
D. 8

3. Which number makes the number sentence true?

$7 \times \square = 0$

A. 0
B. 1
C. 7
D. 14

4. Which is the missing number?

$(22 + 17) + 3 = 22 + (17 + \square)$

A. 42
B. 22
C. 17
D. 3

5. Which number makes the number sentence true?

$\square \times 5 = 5 \times 4$

A. 0
B. 1
C. 4
D. 20

6. Which is the missing number?

$(3 + \square) + 5 = 3 + (4 + 5)$

A. 3
B. 4
C. 5
D. 12

7. What number makes the number sentence true?

$3 \times 8 = \square \times 3$

Answer ____________

8. Omar used 1 pack of stickers to decorate his folder. There are 9 stickers in each pack. How many stickers did Omar use to decorate his folder?

Answer ____________

EXTENDED-RESPONSE QUESTION

9. Sasha multiplied the numbers below.

$11 \times 0 =$ ___

Part A What number makes that number sentence true?

____

Part B What property of multiplication can you use to help you find the product?

____

Part C Sasha has several other numbers to multiply by the product. What will be the products of those multiplication sentences? Explain your answer.

____

____

____

____

Lesson

10 Multiplication Strategies

3.N.19, 3.N.20

Getting the Idea

Here are some strategies you can use to solve multiplication problems.

- Repeated addition
  $6 + 6 + 6 + 6 + 6 + 6 = 36$
- Skip counting
  6, 12, 18, 24, 30, 36
- Multiplication table
  $6 \times 6 = 36$

×	0	1	2	3	4	5	6	7	8	9	10	11	12
0	0	0	0	0	0	0	0	0	0	0	0	0	0
1	0	1	2	3	4	5	6	7	8	9	10	11	12
2	0	2	4	6	8	10	12	14	16	18	20	22	24
3	0	3	6	9	12	15	18	21	24	27	30	33	36
4	0	4	8	12	16	20	24	28	32	36	40	44	48
5	0	5	10	15	20	25	30	35	40	45	50	55	60
6	0	6	12	18	24	30	**36**	42	48	54	60	66	72
7	0	7	14	21	28	35	42	49	56	63	70	77	84
8	0	8	16	24	32	40	48	56	64	72	80	88	96
9	0	9	18	27	36	45	54	63	72	81	90	99	108
10	0	10	20	30	40	50	60	70	80	90	100	110	120
11	0	11	22	33	44	55	66	77	88	99	110	121	132
12	0	12	24	36	48	60	72	84	96	108	120	132	144

- Commutative property of multiplication

  When you know the product of one fact, you know the product of another fact.

  $3 \times 7 = 7 \times 3$

  $21 = 21$

- Double a fact you already know to find a new fact.

  $6 \times 6 = 36$

  $3 \times 6 = 18$

  $3 \times 6 = 18$

EXAMPLE 1

Multiply.

$8 \times 7 = \square$

STRATEGY **Double a known fact.**

STEP 1 One of the factors is 8.

8 is the double of 4.

STEP 2 Think of a known fact: 4×7.

$4 \times 7 = 28$

STEP 3 8 is the double of 4, so double the product of 4×7.

$28 + 28 = 56$, so $8 \times 7 = 56$.

SOLUTION $8 \times 7 = 56$

EXAMPLE 2

Which fact does not have the same product as 3×8?

A. 4×6

B. 8×3

C. 6×4

D. 5×4

STRATEGY **Compare the product of 3×8 to the product for each choice.**

STEP 1 Multiply 3×8.

$3 \times 8 = 24$

Look for the choice that does **not** have a product of 24.

STEP 2 Check Choice A.

$4 \times 6 = 24$

The product is 24. Choice A is not the correct answer.

STEP 3 Check Choice B.

8×3 has the same factors as 3×8.

Both products are 24.

Choice B is not the correct answer.

STEP 4 Check Choice C.

$6 \times 4 = 24$

The product is 24.

Choice C is not the correct answer.

STEP 5 Check Choice D.

$5 \times 4 = 20$

The product is 20.

Choice D is the correct answer.

SOLUTION **Choice D, 5×4, does not have the same product as 3×8.**

Another strategy is to skip count.

EXAMPLE 3

Three groups signed up to hike on a trail. Each group has 7 people. How many people in all are on the trail?

STRATEGY **Use skip counting.**

STEP 1 Show the problem as multiplication.

3 groups of 7 people = 3 groups of 7

$3 \times 7 = ?$

STEP 2 Skip count by 7 three times.

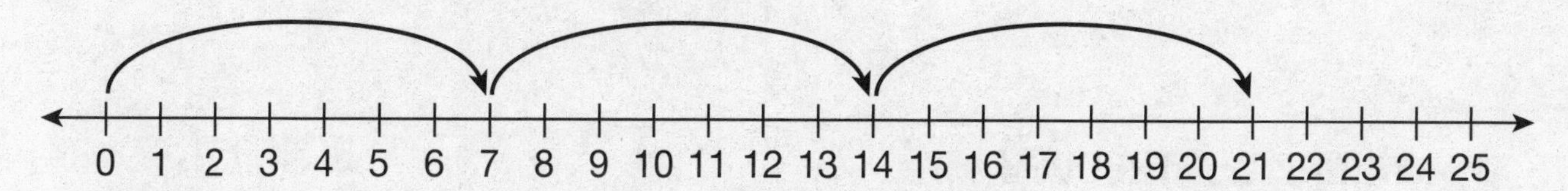

SOLUTION **There are 21 people on the trail in all.**

COACHED EXAMPLE

Write a multiplication fact to show how many ducks in all.

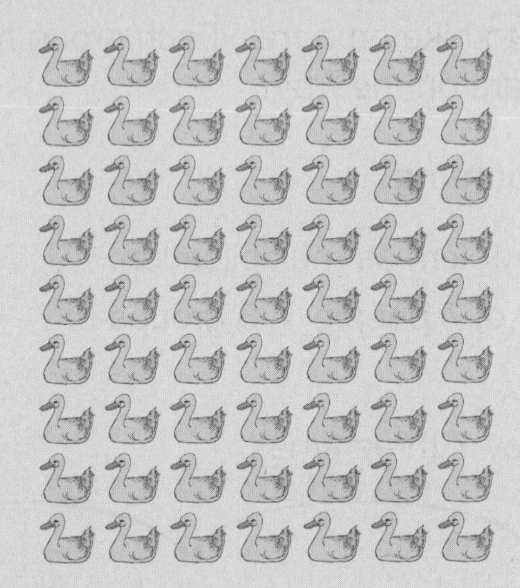

THINKING IT THROUGH

How many rows are there? __________

How many ducks are in each row? __________

Use repeated addition to find how many ducks in all.

____

How many ducks are there in all? _____

Write the multiplication fact.

__________ × __________ = __________

Lesson Practice

Choose the correct answer.

1. $7 \times 2 = \square$

A. 5
B. 9
C. 14
D. 16

2. $6 \times 10 = \square$

A. 6
B. 10
C. 16
D. 60

3. $5 \times 11 = \square$

A. 55
B. 15
C. 11
D. 5

4. $8 \times 8 = \square$

A. 88
B. 64
C. 18
D. 16

5. Which does **not** have the same product as 9×4?

A. 8×7
B. 3×12
C. 6×6
D. 4×9

6. $9 \times 5 = \square$

A. 14
B. 35
C. 40
D. 45

7. $11 \times 2 = \square$

A. 12
B. 21
C. 22
D. 24

8. $4 \times 12 = \square$

A. 16
B. 24
C. 48
D. 64

9. $7 \times 6 = \square$

A. 13
B. 16
C. 23
D. 42

10. Which has the same product as 4×10?

A. 2×5
B. 5×8
C. 7×6
D. 9×9

11. Millie baked 5 dozen cookies. There are 12 cookies in a dozen. How many cookies did Millie bake in all?

Answer __________

12. Mr. Field's garden has 8 rows of plants. Each row has 10 plants. How many plants does Mr. Field's garden have in all?

Answer __________

EXTENDED-RESPONSE QUESTION

13. Steven bought 6 bags of potatoes. Each bag has 6 potatoes.

Part A Use the space below to make a model of the problem.

Part B How many potatoes did Steven buy?

____

Part C Explain the strategy you used to find the answer to Part B.

____

____

____

____

Lesson

11 Division Facts

Getting the Idea

When you **divide**, you separate objects into equal groups, or find the number of objects in each group.

Here are the parts to a division sentence.

6	÷	3	=	2
dividend		**divisor**		**quotient**

You can use an array to find the number of equal groups.

EXAMPLE 1

Casey had 32 dimes. He gave 8 dimes to each friend. To how many friends did Casey give dimes?

STRATEGY **Make an array.**

STEP 1 Make equal rows.

Use 32 dimes. Put 8 dimes in each row.

STEP 2 Count the number of rows.

There are 4 rows.

STEP 3 Write a division sentence.

32	÷	8	=	4
total dimes		number of dimes for each friend		number of friends receiving dimes

SOLUTION **Casey gave dimes to 4 friends.**

You can use a multiplication table to help you with division facts.

EXAMPLE 2

18 ÷ 2 = □

STRATEGY **Use a multiplication table.**

×	0	1	2	3	4	5	6	7	8	9	10
0	0	0	0	0	0	0	0	0	0	0	0
1	0	1	2	3	4	5	6	7	8	9	10
2	0	2	4	6	8	10	12	14	16	18	20
3	0	3	6	9	12	15	18	21	24	27	30
4	0	4	8	12	16	20	24	28	32	36	40
5	0	5	10	15	20	25	30	35	40	45	50
6	0	6	12	18	24	30	36	42	48	54	60
7	0	7	14	21	28	35	42	49	56	63	70
8	0	8	16	24	32	40	48	56	64	72	80
9	0	9	18	27	36	45	54	63	72	81	90
10	0	10	20	30	40	50	60	70	80	90	100

STEP 1 Look at the 2s row. Find 18.

STEP 2 From 18, go to the top of the column. Find the number there. It is 9. The quotient is 9.

SOLUTION **18 ÷ 2 = 9**

You can use repeated subtraction to find the quotient.

EXAMPLE 3

$12 \div 3 = \square$

STRATEGY **Use repeated subtraction.**

STEP 1 Start with 12. Subtract 3 until you reach 0.

$12 - 3 = 9$

$9 - 3 = 6$

$6 - 3 = 3$

$3 - 3 = 0$

STEP 2 Count the number of times you subtracted 3.

You subtracted 4 times.

SOLUTION $12 \div 3 = 4$

Multiplication and division are **inverse operations** or opposites.

Inverse operations undo each other. So you can use a multiplication fact to solve a division fact, or a division fact to solve a multiplication fact.

A **fact family** is a group of related facts that use the same numbers. Here is the fact family for 2, 3, and 6.

$3 \times 2 = 6$ $2 \times 3 = 6$

$6 \div 3 = 2$ $6 \div 2 = 3$

EXAMPLE 4

Mr. Frey has 24 students. He seated the students at 4 tables. Each table has the same number of students. How many students are at each table?

STRATEGY **Write a division sentence. Use a related multiplication fact.**

STEP 1 Write the division sentence for the problem.

24 students in 4 equal groups

$24 \div 4 = ?$

STEP 2 Think about a related multiplication fact.

$4 \times ? = 24$

$4 \times 6 = 24$

STEP 3 Multiplication and division are inverse operations.

$4 \times 6 = 24$

$24 \div 4 = 6$

SOLUTION **There are 6 students at each table.**

COACHED EXAMPLE

Write a division sentence for this picture.

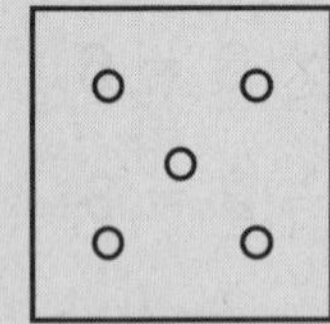
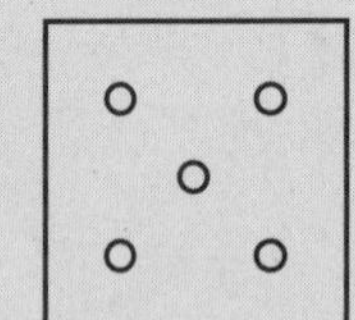
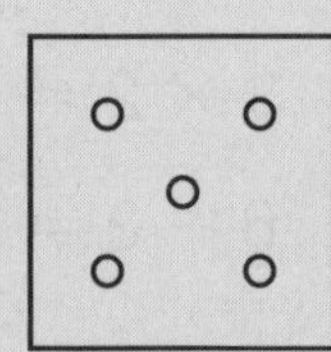

THINKING IT THROUGH

How many counters are there in all? __________

How many equal groups of counters are there? _____

How many counters are in each group? __________

Write the division sentence. ________ ÷ ________ = ________

The picture shows the division sentence ____________________.

Lesson Practice

Choose the correct answer.

1. Which division fact does the picture show?

A. $6 \times 3 = 18$

B. $12 \div 3 = 4$

C. $15 \div 3 = 5$

D. $18 \div 3 = 6$

2. Four students share 20 marbles. Each student gets the same number of marbles. How many marbles does each student get?

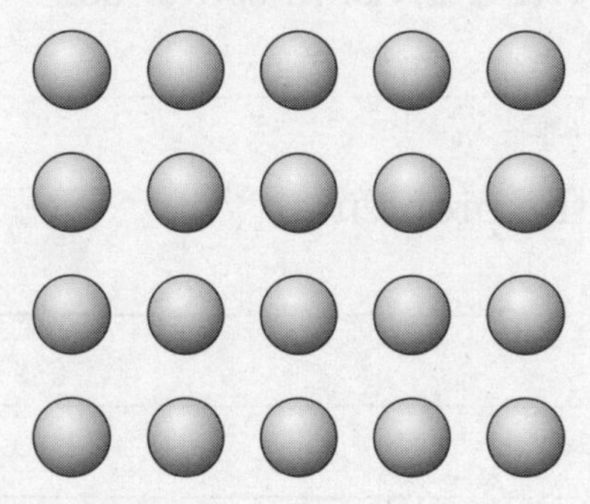

A. 3

B. 4

C. 5

D. 6

3. Which division fact is related to this multiplication fact?

$$2 \times 4 = 8$$

A. $8 \div 1 = 8$

B. $4 \div 2 = 2$

C. $8 \div 4 = 2$

D. $4 \div 4 = 1$

4. Divide.

$$21 \div 3 = \square$$

A. 6

B. 7

C. 8

D. 9

5. Which does **not** have a quotient of 6?

A. $30 \div 5 = \square$

B. $24 \div 4 = \square$

C. $18 \div 3 = \square$

D. $14 \div 2 = \square$

6. Divide.

$24 \div 3 = \square$

Answer ____________

7. There are 25 students that signed up for a clean-up project. They formed teams of 5 students. How many teams are there?

Answer ____________

EXTENDED-RESPONSE QUESTION

8. Mrs. Martinez gives her three children $21 to share equally.

Part A Draw a picture to show the problem.

Part B How much did each child receive? Write a division sentence.

____

Part C Explain how you solved each part of the problem.

____

____

____

12 Division Strategies

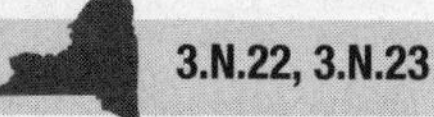

Getting the Idea

Here are some strategies to use to solve division problems.

- Draw a picture or make a model.
- Use a related division fact or multiplication fact.
- Skip count backward.
- Use repeated subtraction.

EXAMPLE 1

Ralph bought 20 stamps. He will give half of the stamps to Kay. How many stamps will Ralph give to Kay?

STRATEGY **Draw a picture.**

STEP 1 Decide what you need to do.

"Half of the stamps" means to divide by 2.

Divide 20 into 2 equal groups.

STEP 2 Draw a picture.

Make 20 circles. Divide into 2 equal groups.

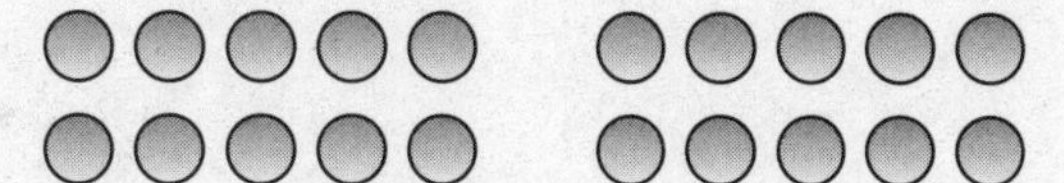

STEP 3 Count how many are in each group.

There are 10 circles in each group.

SOLUTION **Ralph will give 10 stamps to Kay.**

EXAMPLE 2

Seven children share 56 grapes. Each child gets the same number of grapes. How many grapes does each child get?

STRATEGY **Use repeated subtraction.**

STEP 1 Decide what you need to do.

You need to divide 56 into 7 equal groups.

Divide $56 \div 7$.

STEP 2 Use repeated subtraction.

Start with 56. Subtract 7 until you reach 0.

$56 - 7 = 49$

$49 - 7 = 42$

$42 - 7 = 35$

$35 - 7 = 28$

$28 - 7 = 21$

$21 - 7 = 14$

$14 - 7 = 7$

$7 - 7 = 0$

STEP 3 Count the number of times you subtracted 7.

You subtracted 8 times.

$56 \div 7 = 8$

SOLUTION **Each child gets 8 grapes.**

You can also use skip counting to solve division problems. Start from the total and skip count backward until you reach 0.

EXAMPLE 3

Katarina wants to share 15 flowers equally among 5 friends. How many flowers should each friend receive?

STRATEGY **Skip count backward.**

STEP 1 Decide what you need to do.

Share 15 flowers among 5 friends.

Divide 15 ÷ 5.

STEP 2 Find the total and the number of groups.

The total is 15. The number of groups is 5.

STEP 3 Skip count backward by 5s from 15 to 0.

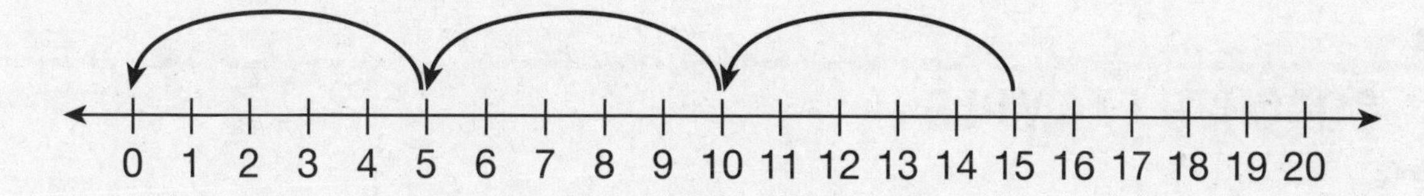

STEP 4 Find how many skips were made.

There are 3 skips.

15 ÷ 5 = 3

SOLUTION **Each friend will receive 3 flowers.**

Remember, a fact family has related facts that use the same numbers.

EXAMPLE 4

$40 \div 8 = \square$

STRATEGY **Use a related multiplication fact.**

STEP 1 Look at the numbers in the problem.

40 and 8

STEP 2 Think of a related multiplication fact.

$8 \times ? = 40$ $8 \times 5 = 40$

STEP 3 Multiplication and division are inverse operations.

$8 \times 5 = 40$ So, $40 \div 8 = 5$

SOLUTION $\mathbf{40 \div 8 = 5}$

COACHED EXAMPLE

Nick has 72 DVDs in his collection. He puts 6 DVDs in each page of an album. How many pages of the album did Nick use for his DVDs?

THINKING IT THROUGH

Decide what you need to do.

You need to divide 72 into 6 equal groups.

Divide _____ ÷ _____.

What related multiplication fact uses the numbers 72 and 6?

_____ × _____ = _____

So, _____ ÷ _____ = _____.

Nick used _______ pages of the album for his DVDs.

Lesson Practice

Choose the correct answer.

1. $32 \div 4 = \square$

A. 16
B. 12
C. 9
D. 8

2. $70 \div 7 = \square$

A. 7
B. 10
C. 63
D. 77

3. $81 \div 9 = \square$

A. 6
B. 7
C. 8
D. 9

4. $33 \div 3 = \square$

A. 10
B. 11
C. 30
D. 33

5. $28 \div 7 = \square$

A. 11
B. 6
C. 4
D. 1

6. $90 \div 10 = \square$

A. 8
B. 9
C. 10
D. 100

7. $42 \div 7 = \square$

A. 5
B. 6
C. 8
D. 49

8. $45 \div 5 = \square$

A. 8
B. 9
C. 10
D. 40

9. $72 \div 8 = \square$

A. 9

B. 10

C. 12

D. 80

10. $24 \div 2 = \square$

A. 26

B. 22

C. 12

D. 10

11. Which is a related multiplication fact to this division fact?

$$48 \div 8 = 6$$

A. $4 \times 12 = 48$

B. $48 - 6 = 42$

C. $8 \times 8 = 64$

D. $6 \times 8 = 48$

12. A chess club has 16 members. Half of the members attended a tournament. How many members attended the tournament?

Answer __________

13. Mark plans to read 63 pages of his book this week. If Mark reads the same number of pages each day, how many pages will he read each day? (Hint: There are 7 days in a week.)

Answer __________

13 Unit Fractions

3.N.11, 3.N.12, 3.N.13

Getting the Idea

A **fraction** names part of a whole or part of a group.

The **numerator** is the top number in a fraction.

The **denominator** is the bottom number in a fraction. It tells the total number of equal parts that make up the whole.

Part of a Whole

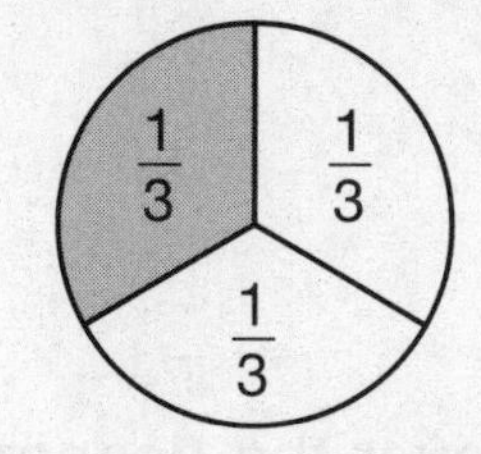

The circle shows 3 equal parts.

Each part is $\frac{1}{3}$ of the circle.

$$\frac{\text{number of shaded parts}}{\text{total number of equal parts}} \begin{matrix} \rightarrow \\ \rightarrow \end{matrix} \frac{1}{3}$$

The fraction $\frac{1}{3}$ is read *one-third*.

Part of a Group

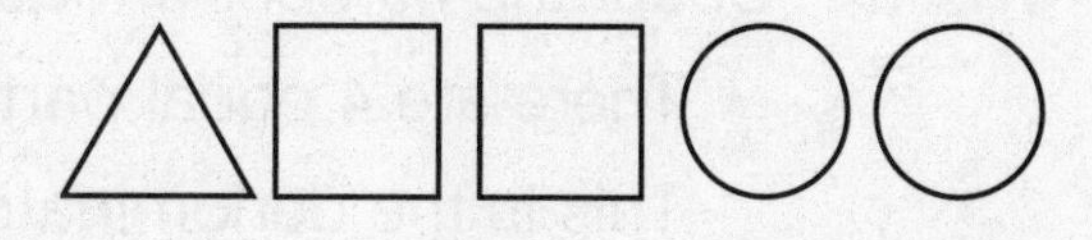

The picture shows 5 shapes.

One of the shapes is a triangle.

The triangle is $\frac{1}{5}$ of the shapes.

$$\frac{\text{number of triangles}}{\text{total number of shapes}} \begin{matrix} \rightarrow \\ \rightarrow \end{matrix} \frac{1}{5}$$

The fraction $\frac{1}{5}$ is read *one-fifth*.

A fraction with 1 as its numerator is called a **unit fraction**.

EXAMPLE 1

What fraction of the figure is shaded? Is the fraction a unit fraction?

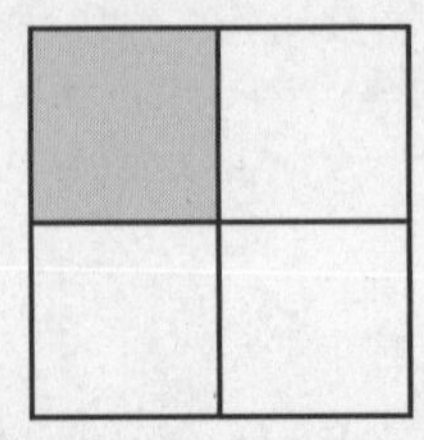

STRATEGY **Find the number of shaded parts, and the number of equal parts in all. Then decide if the fraction is a unit fraction.**

STEP 1 Count the number of shaded parts.

There is 1 shaded part.

This is the numerator.

STEP 2 Count the number of equal parts.

There are 4 equal parts.

This is the denominator.

STEP 3 Write the fraction. Write the numerator over the denominator.

$\frac{1}{4}$

STEP 4 Decide if the fraction is a unit fraction.

The numerator is 1.

It is a unit fraction.

SOLUTION **The figure is $\frac{1}{4}$ shaded. Yes, $\frac{1}{4}$ is a unit fraction.**

EXAMPLE 2

What fraction of the stars are gray? How would you read that fraction?

STRATEGY **Find the numerator and the denominator.**

STEP 1 Count the number of gray stars.

There is 1 gray star.

This is the numerator.

STEP 2 Count the total number of stars.

There are 6 total stars.

This is the denominator.

STEP 3 Write the fraction.

$\frac{\text{numerator}}{\text{denominator}} = \frac{1}{6}$

STEP 4 Decide how the fraction should be read.

The fraction is $\frac{1}{6}$. It should be read *one-sixth*.

SOLUTION **$\frac{1}{6}$ of the stars are gray. The fraction should be read *one-sixth*.**

COACHED EXAMPLE

What fraction of the rectangle is shaded?
Is the fraction a unit fraction?

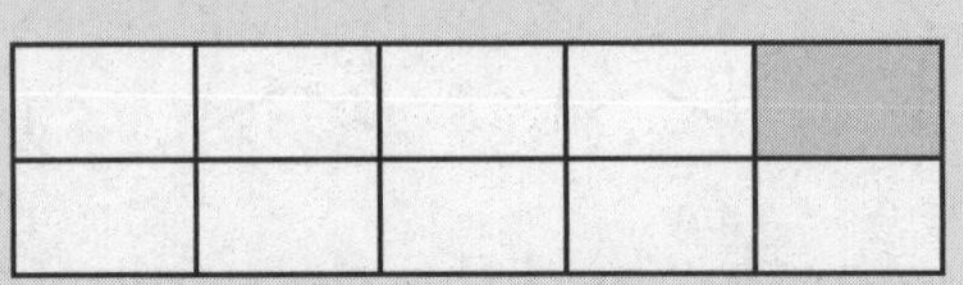

THINKING IT THROUGH

How many parts are shaded? _____

What is the numerator? _____

How many equal parts are in the rectangle? _____

What is the denominator? _____

Write the fraction.

$\frac{\text{numerator}}{\text{denominator}} = \frac{\square}{\square}$

__________ of the rectangle is shaded.

Since the numerator is __________, it __________ a unit fraction.

Lesson Practice

Choose the correct answer.

1. What fraction of the circle is shaded?

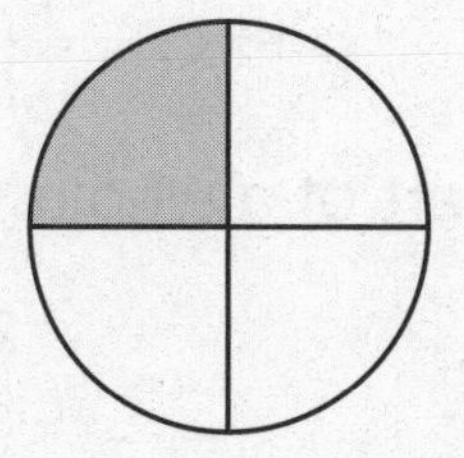

A. $\frac{1}{3}$ **C.** $\frac{1}{5}$

B. $\frac{1}{4}$ **D.** $\frac{1}{6}$

2. What fraction of the rectangle is shaded?

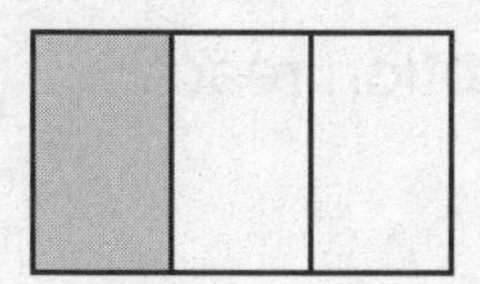

A. $\frac{1}{2}$ **C.** $\frac{3}{1}$

B. $\frac{1}{3}$ **D.** $\frac{2}{1}$

3. How can you read this unit fraction?

$$\frac{1}{2}$$

A. one-half

B. one-third

C. one-two

D. two-one

4. What fraction of the arrows is white?

A. $\frac{1}{3}$ **C.** $\frac{3}{1}$

B. $\frac{1}{4}$ **D.** $\frac{4}{1}$

5. What fraction of the pentagon is gray?

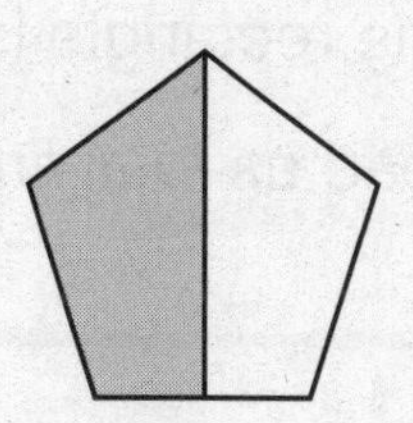

Answer ____________

6. What fraction of the shapes is the circle?

Answer ____________

14 Models for Fractions

3.N.10, 3.N.11, 3.N.12, 3.N.13

Getting the Idea

Remember, fractions are numbers that name part of a whole.

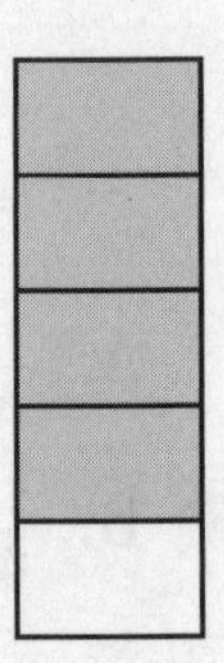

The rectangle to the right has 5 equal parts.

Each part is $\frac{1}{5}$ of the rectangle.

4 parts are shaded.

$\frac{4}{5}$ of the rectangle is shaded.

It is read as *four-fifths.*

EXAMPLE 1

What part of the hexagon is shaded? How is that fraction read?

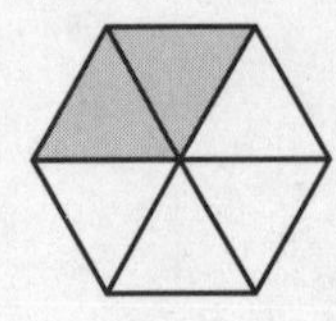

STRATEGY **Find the denominator and the numerator.**

STEP 1 Count the total number of parts.

There are 6. This is the denominator.

STEP 2 Count the number of shaded parts.

There are 2. This is the numerator.

STEP 3 Write the fraction.

$$\frac{\text{numerator}}{\text{denominator}} = \frac{2}{6}$$

SOLUTION **$\frac{2}{6}$ of the hexagon is shaded. It is read as *two-sixths.***

EXAMPLE 2

Make a model to show $\frac{3}{4}$.

STRATEGY **Look at the denominator and the numerator.**

STEP 1 Look at the denominator.

The 4 represents the number of equal parts.

STEP 2 Make the drawing.

Draw a rectangle. Make 4 equal parts.

STEP 3 Look at the numerator.

The 3 represents the number of parts to shade.

STEP 4 Shade 3 parts of the rectangle.

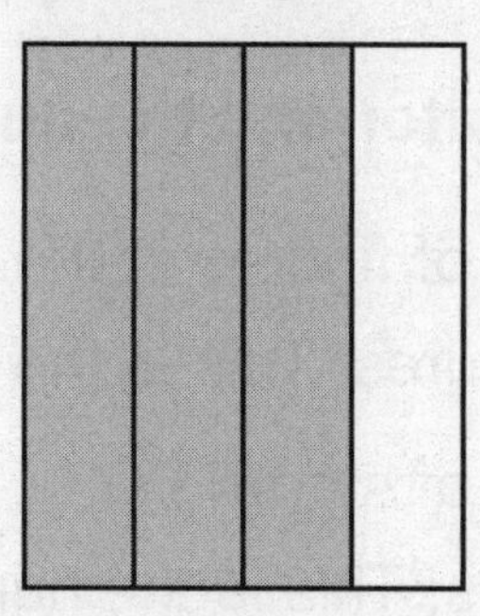

SOLUTION **The model shows $\frac{3}{4}$ of the rectangle is shaded.**

Remember, a fraction can also name part of a group.

There are 5 children in the group.

Each child is $\frac{1}{5}$ of the group.

2 of the children are girls.

Girls are $\frac{2}{5}$ of the group.

The fraction $\frac{2}{5}$ is read as *two-fifths*.

EXAMPLE 3

What fraction of the figures is stars? How is that fraction read?

STRATEGY **Find the denominator and the numerator.**

STEP 1 Count the number of figures in the group.

There are 10 figures. This is the denominator.

STEP 2 Count the number of stars.

There are 4 stars. This is the numerator.

STEP 3 Write the fraction.

$$\frac{\text{numerator}}{\text{denominator}} = \frac{4}{10}$$

SOLUTION **$\frac{4}{10}$ of the total number of figures is stars.**

It is read as *four-tenths.*

COACHED EXAMPLE

What fraction of the circle is shaded?

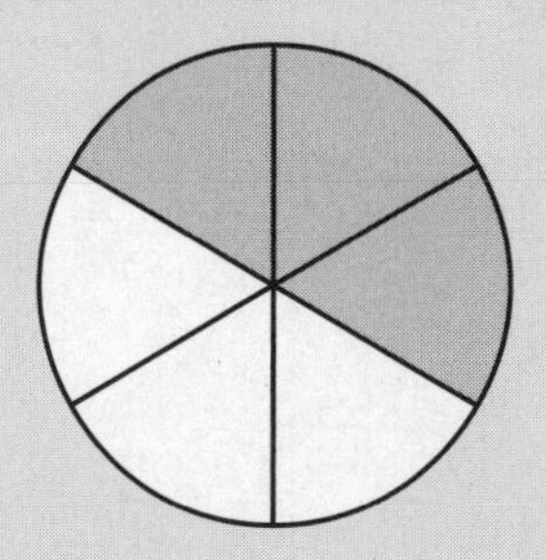

THINKING IT THROUGH

How many equal parts are there? _____

What is the denominator? _____

How many parts are shaded? _____

What is the numerator? _____

Write the fraction.

$\frac{\text{numerator}}{\text{denominator}} = \frac{\square}{\square}$

So, _______ of the circle is shaded.

Lesson Practice

Choose the correct answer.

1. What fraction of the rectangle is shaded?

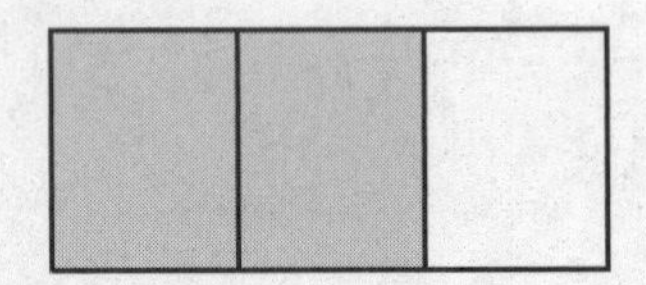

A. $\frac{1}{3}$

B. $\frac{2}{3}$

C. $\frac{1}{2}$

D. $\frac{1}{4}$

2. Which figure shows $\frac{2}{4}$ shaded?

A.

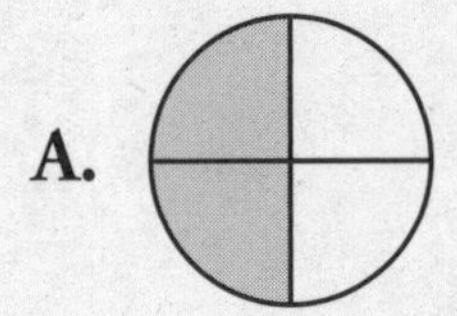

B.

C.

D.

3. What fraction of the total number of shapes is hearts?

A. $\frac{3}{5}$ **C.** $\frac{3}{4}$

B. $\frac{2}{5}$ **D.** $\frac{2}{3}$

4. What fraction of the total number of fruits is apples?

A. $\frac{3}{4}$ **C.** $\frac{4}{7}$

B. $\frac{3}{7}$ **D.** $\frac{2}{4}$

5. Which fraction is read two-ninths?

A. $\frac{1}{2}$

B. $\frac{9}{2}$

C. $\frac{1}{7}$

D. $\frac{2}{9}$

6. What fraction of the circle is shaded?

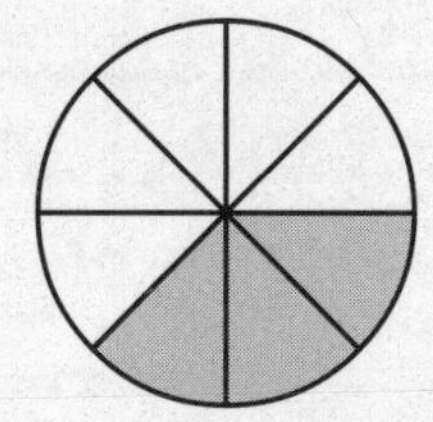

Answer ____________

7. What fraction of the total number of figures are suns?

Answer ____________

EXTENDED-RESPONSE QUESTION

8. Lenny wants to shade a model to show $\frac{5}{8}$.

Part A Shade the rectangle below to show $\frac{5}{8}$.

Part B In the space below, draw a model to show $\frac{5}{8}$ of a group.

Part C Explain how you solved each part of the problem.

____

____

____

____

Lesson

15 Explore Equivalent Fractions

Getting the Idea

Fractions that name the same parts of a whole are called **equivalent fractions.**

EXAMPLE 1

Write two equivalent fractions that name the shaded parts of the circle.

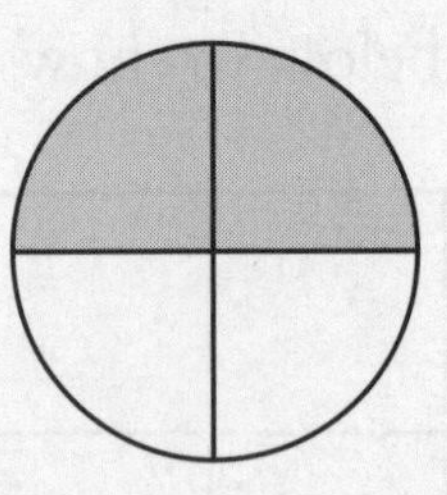

STRATEGY **Look at the shaded parts of the circle in two ways.**

STEP 1 Count the number of equal parts and the number of shaded parts.

There are 4 equal parts. There are 2 shaded parts.

So, $\frac{2}{4}$ of the circle is shaded.

STEP 2 Look at the shaded parts another way.

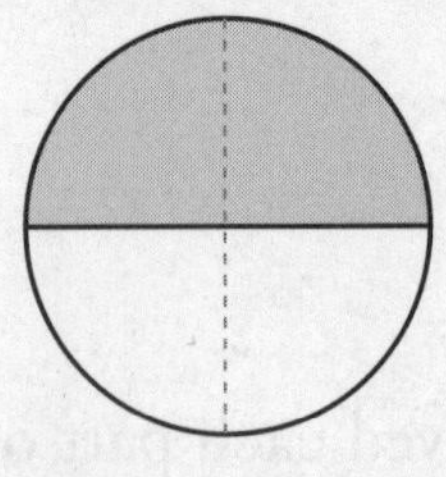

The two shaded parts are half of the circle.

One half of the circle is shaded.

So, $\frac{1}{2}$ of the circle is shaded.

SOLUTION **The fractions $\frac{1}{2}$ and $\frac{2}{4}$ name the shaded parts of the circle.**

EXAMPLE 2

Are $\frac{1}{2}$ and $\frac{4}{8}$ equivalent fractions?

STRATEGY **Use fraction strips.**

STEP 1 Use the fraction strips for $\frac{1}{2}$ and $\frac{1}{8}$.

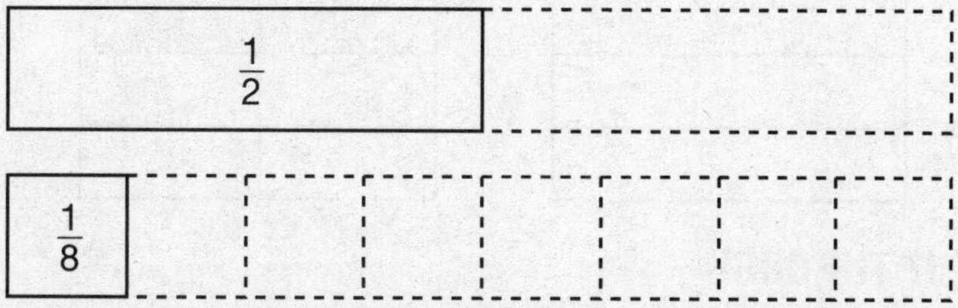

$\frac{1}{2}$

$\frac{1}{8}$

STEP 2 Put together four $\frac{1}{8}$ strips to equal the length of the $\frac{1}{2}$ strip.

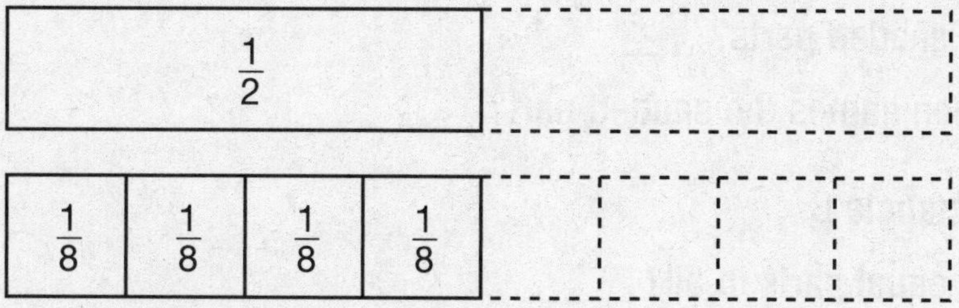

$\frac{1}{2}$

$\frac{1}{8}$ $\frac{1}{8}$ $\frac{1}{8}$ $\frac{1}{8}$

They are the same length.

The two fractions are equivalent.

$\frac{1}{2} = \frac{4}{8}$

SOLUTION **Yes, $\frac{1}{2}$ and $\frac{4}{8}$ are equivalent fractions.**

COACHED EXAMPLE

Write the fraction that names the shaded parts in each rectangle.
Are the two fractions equivalent?

A B

THINKING IT THROUGH

Look at rectangle A.

How many equal parts in all? _____

How many shaded parts? _____

What fraction names the shaded part? ______

Look at rectangle B.

How many equal parts in all? _____

How many shaded parts? _____

What fraction names the shaded parts? _____

Are the shaded areas in the rectangles the same size? ___________

Are the fractions equivalent? ___________

Rectangle A shows ________ of the rectangle shaded.

Rectangle B shows ________ of the rectangle shaded.

The two fractions ________ and ________ are equivalent.

Lesson Practice

Choose the correct answer.

1. Which fraction is equivalent to $\frac{1}{4}$?

A. $\frac{1}{8}$

B. $\frac{2}{8}$

C. $\frac{4}{8}$

D. $\frac{6}{8}$

2. Which fraction is equivalent to $\frac{3}{9}$?

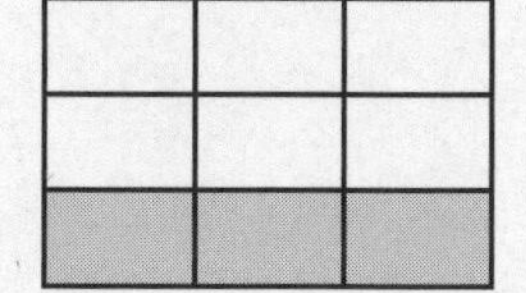

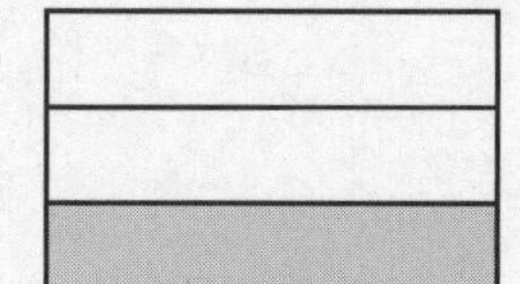

A. $\frac{1}{9}$

B. $\frac{1}{3}$

C. $\frac{1}{4}$

D. $\frac{1}{2}$

3. Look at the circle below.

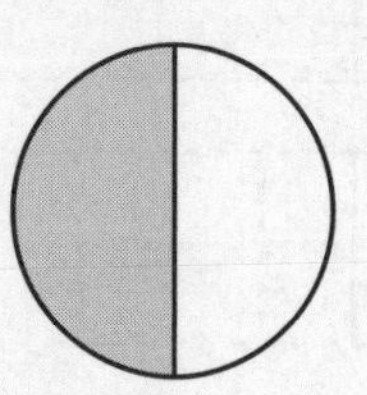

Which shows $\frac{1}{2}$ of the circle shaded?

A. 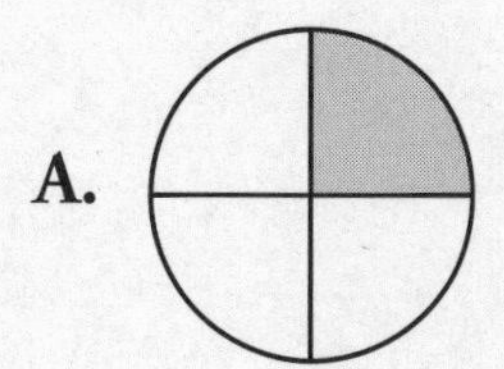

B. 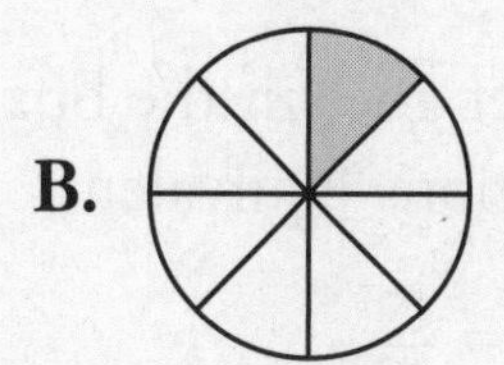

C. 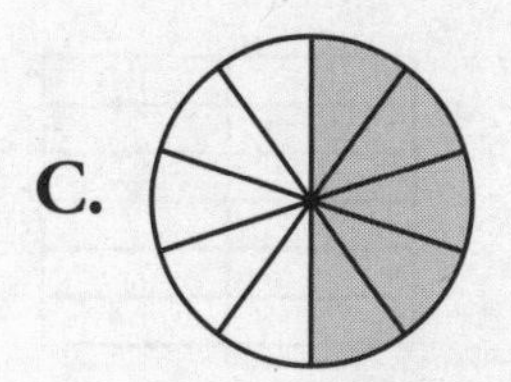

D. 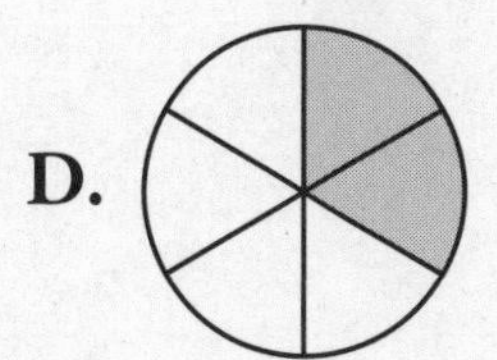

4. Which fraction is equivalent to $\frac{1}{3}$?

$\frac{1}{3}$

$\frac{1}{12}$ $\frac{1}{12}$ $\frac{1}{12}$ $\frac{1}{12}$

A. $\frac{1}{12}$

B. $\frac{4}{12}$

C. $\frac{1}{4}$

D. $\frac{4}{8}$

5. Which number goes in the box to make the fractions equivalent?

$$\frac{1}{4} = \frac{\square}{12}$$

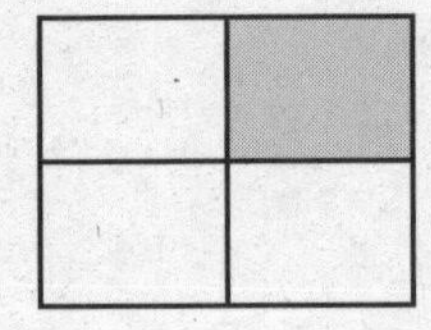

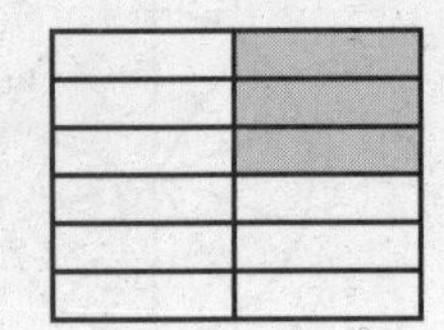

A. 1

B. 3

C. 6

D. 12

6. Look at the models below.

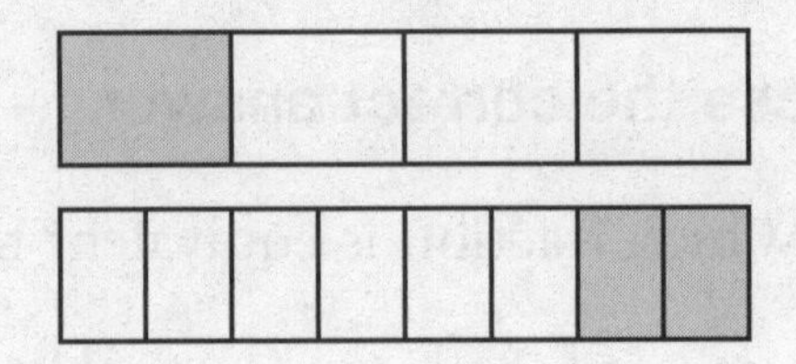

What fraction is equal to $\frac{1}{4}$?

Answer ______________

7. Write two equivalent fractions for the models below.

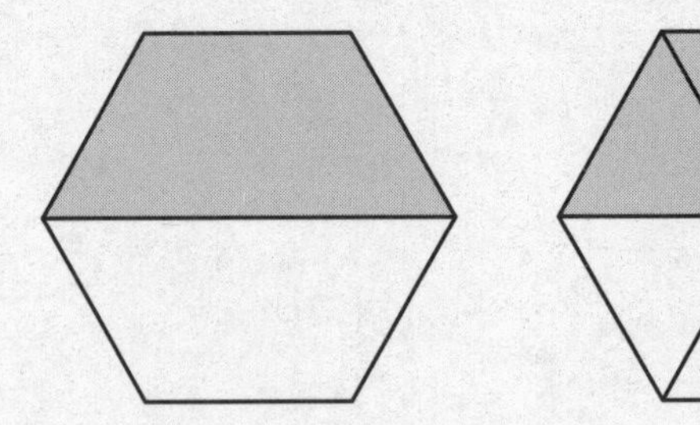

Answer ______________

EXTENDED-RESPONSE QUESTION

8. Vanessa and Niles each have a pizza that is the same size. Vanessa cut her pizza into 3 equal pieces. She ate 1 piece. Niles cut his pizza into 6 equal pieces. Niles ate the same amount of pizza as Vanessa.

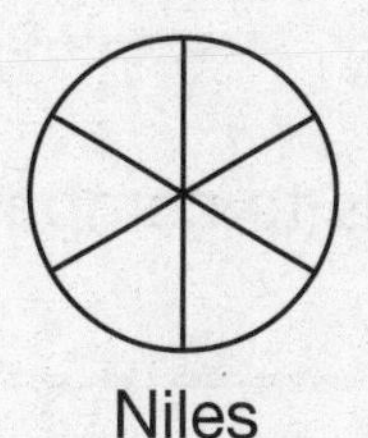

Part A Shade the circle to show the amount of pizza Vanessa ate.

Part B Shade the circle to show the amount of pizza Niles ate.

Part C How many pieces of pizza did Niles eat? Explain how you found your answer.

____

____

____

____

Lesson

16 Compare and Order Unit Fractions

Getting the Idea

You can use models to compare fractions.

EXAMPLE 1

Which is the greater fraction, $\frac{1}{3}$ or $\frac{1}{4}$?

STRATEGY **Make a drawing to show each fraction.**

STEP 1 Draw rectangles to represent each fraction.

One rectangle has 3 equal parts. 1 part is shaded.

Another rectangle has 4 equal parts. 1 part is shaded.

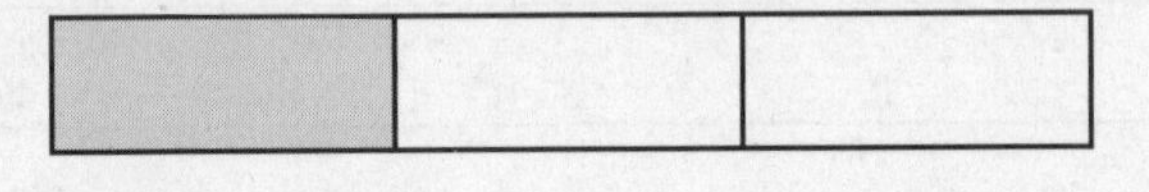

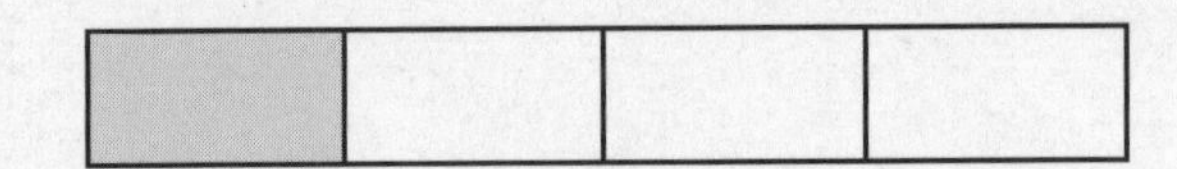

STEP 2 Compare the shaded parts.

The shaded area that shows $\frac{1}{3}$ is bigger than the shaded area that shows $\frac{1}{4}$.

SOLUTION **$\frac{1}{3}$ is the greater fraction.**

You can use number lines to help you compare and order fractions.

EXAMPLE 2

Order $\frac{1}{2}$, $\frac{1}{8}$, and $\frac{1}{4}$ from greatest to least.

STRATEGY **Use number lines to order the fractions.**

STEP 1 Draw number lines divided into halves, eighths, and fourths.

Make them all the same length, and line them up on 0.

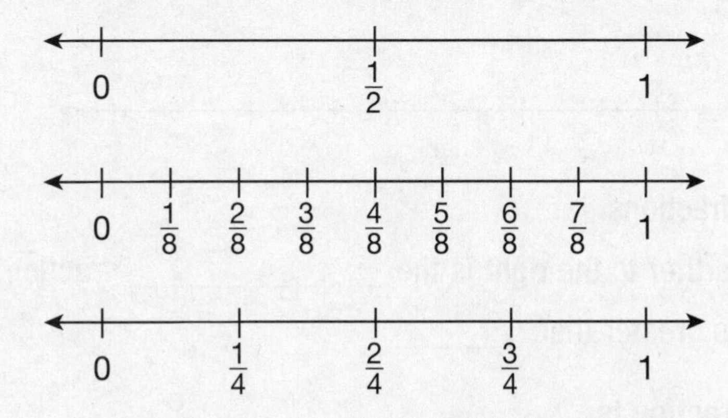

STEP 2 Place points on the number lines at $\frac{1}{2}$, $\frac{1}{8}$, and $\frac{1}{4}$.

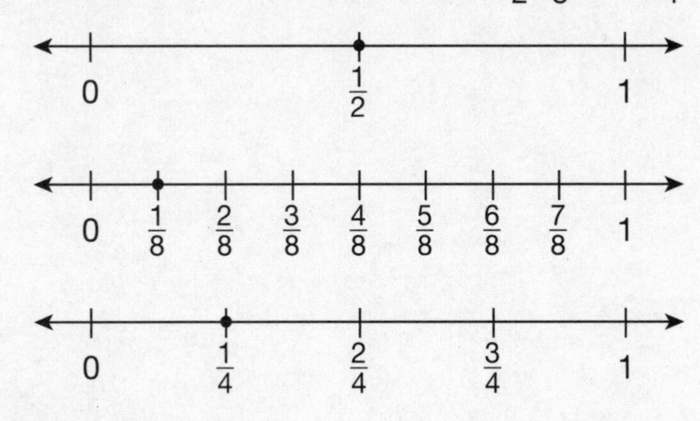

STEP 3 Order the fractions.

The fraction farthest to the right is the greatest fraction.

$\frac{1}{2}$ is the greatest fraction.

The fraction farthest to the left is the least fraction.

$\frac{1}{8}$ is the least fraction.

SOLUTION **The fractions from greatest to least are $\frac{1}{2}$, $\frac{1}{4}$, and $\frac{1}{8}$.**

COACHED EXAMPLE

Which is the greater fraction, $\frac{1}{2}$ or $\frac{1}{4}$?

THINKING IT THROUGH

Draw number lines divided into halves and fourths.

Place points at $\frac{1}{2}$ and $\frac{1}{4}$ on the number lines.

Compare the fractions.

The fraction farther to the right is the ______________ fraction.

So _______ is greater than _______.

The greater fraction is _____.

Lesson Practice

Choose the correct answer.

1. Look at these two fractions.

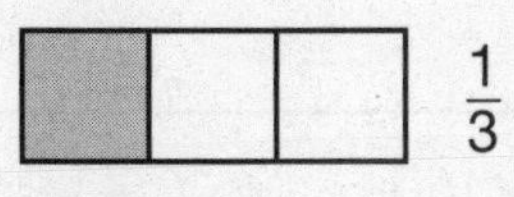
$\frac{1}{3}$

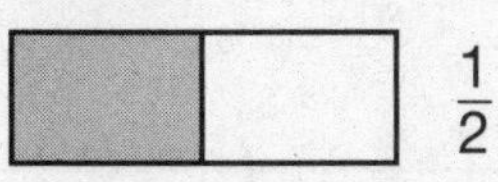
$\frac{1}{2}$

Which sentence is true?

A. $\frac{1}{2}$ is less than $\frac{1}{3}$.

B. $\frac{1}{2}$ is greater than $\frac{1}{3}$.

C. $\frac{1}{3}$ is equal to $\frac{1}{2}$.

D. $\frac{1}{3}$ is greater than $\frac{1}{2}$.

2. Look at these two fractions.

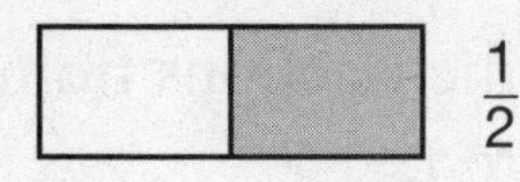
$\frac{1}{2}$

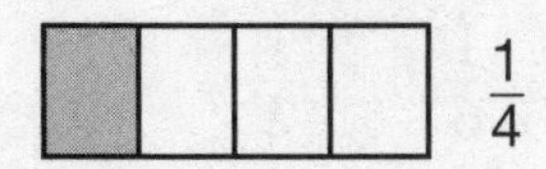
$\frac{1}{4}$

Which sentence is true?

A. $\frac{1}{2}$ is less than $\frac{1}{4}$.

B. $\frac{1}{2}$ is equal to $\frac{1}{4}$.

C. $\frac{1}{4}$ is greater than $\frac{1}{2}$.

D. $\frac{1}{4}$ is less than $\frac{1}{2}$.

3. Which fraction is the least?

A. 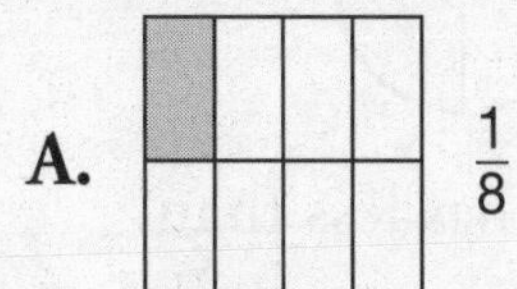 $\frac{1}{8}$

B. 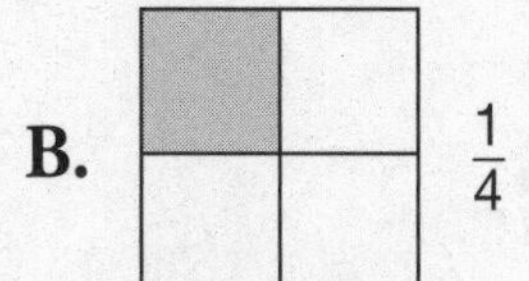 $\frac{1}{4}$

C. 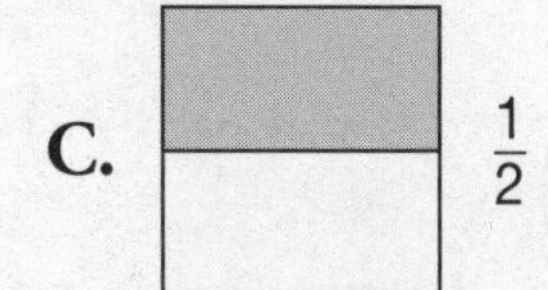 $\frac{1}{2}$

D. 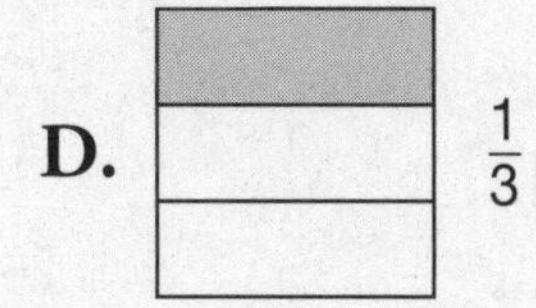 $\frac{1}{3}$

4. Which fraction is the greatest?

A. $\frac{1}{8}$

B. $\frac{1}{4}$

C. $\frac{1}{2}$

D. $\frac{1}{3}$

5. This circle is $\frac{1}{4}$ shaded.

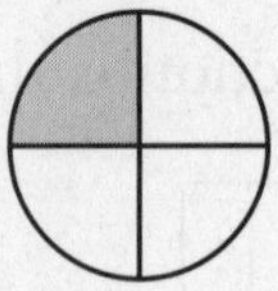

Which circle has less than $\frac{1}{4}$ shaded?

A. 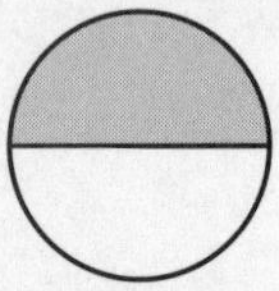

B. 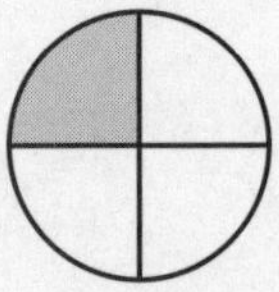

C. 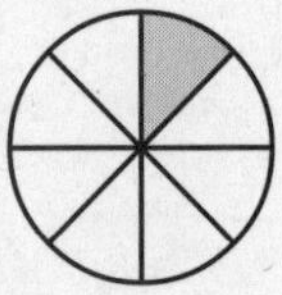

D. 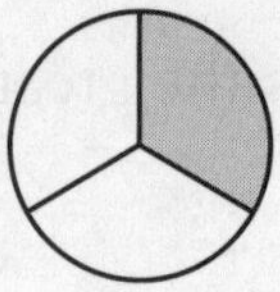

6. Which fraction is the least?

A. $\frac{1}{2}$

B. $\frac{1}{3}$

C. $\frac{1}{4}$

D. $\frac{1}{6}$

Use the number lines for questions 7 and 8.

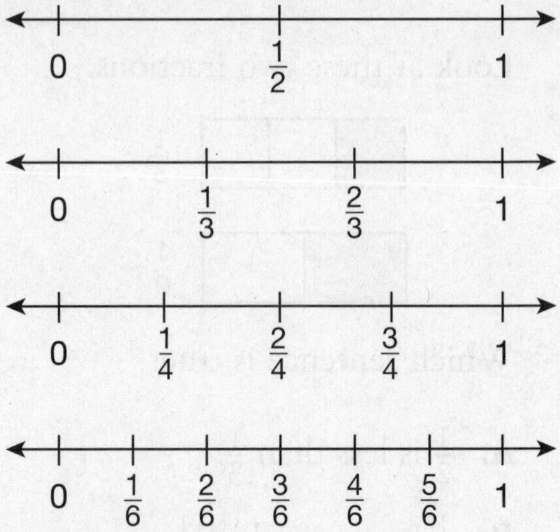

7. Order the fractions from greatest to least.

$\frac{1}{4}$ $\frac{1}{6}$ $\frac{1}{3}$

Answer ____________________

8. Order the fractions from least to greatest.

$\frac{1}{6}$ $\frac{1}{2}$ $\frac{1}{8}$

Answer ____________________

Lesson

17 Solve Word Problems

Getting the Idea

When you solve a word problem, first identify the information given in the problem. Key words tell you which **operation** to use.

Key Words

Addition	Subtraction	Multiplication	Division
total	how many more	total	half
in all	how many fewer	in all	how many groups
combined	how many left		how many in each

Once you know the operation to use, decide how to solve the problem. Use paper and pencil, mental math, estimation, or a calculator.

EXAMPLE 1

Naima baked 280 cookies for a fundraiser. She sold 215 cookies. How many cookies did Naima have left?

STRATEGY **Look for key words to decide which operation to use.**

STEP 1 Identify the information.

Naima baked 280 cookies. She sold 215 cookies.

STEP 2 Decide which operation to use. Look for key words.

"How many left" tells you to subtract.

STEP 3 Use paper and pencil to subtract.

$$\begin{array}{r} {\scriptstyle 7\,10} \\ 2\,\not8\,\not0 \\ -\,2\,1\,5 \\ \hline 6\,5 \end{array}$$

SOLUTION **Naima had 65 cookies left.**

EXAMPLE 2

Peter put 5 place settings on the dinner table. He put 2 forks at each setting. How many forks did Peter put on the dinner table in all?

STRATEGY **Decide which operation to use.**

STEP 1 Identify the information.

Peter put 5 place settings. There are 2 forks at each setting.

STEP 2 Decide which operation to use. Look for key words.

"How many forks in all?" tells you to add or multiply.

Since this is in equal groups ("2 forks at each setting"), use multiplication.

STEP 3 Multiply.

5 groups of 2

$5 \times 2 = 10$

SOLUTION **Peter put 10 forks on the dinner table.**

EXAMPLE 3

Tina collected 82 shells at the beach. José collected 88 shells at the beach. About how many shells did they collect in all?

STRATEGY **Decide which operation to use.**

STEP 1 Identify the information.

Tina collected 82 shells. José collected 88 shells.

STEP 2 Decide which operation to use. Look for key words.

"How many in all?" tells you to add.

The word "about" tells you to estimate.

STEP 3 Use mental math to add.

Round each number to the nearest 10.

82 rounds to 80. 88 rounds to 90.

$80 + 90 = 170$

SOLUTION **They collected about 170 shells in all.**

EXAMPLE 4

Clint had 30 markers. He separated them into 5 equal groups. How many markers are in each group?

STRATEGY **Decide which operation to use.**

STEP 1 Identify the information.

Clint had 30 markers. He put them into 5 groups.

STEP 2 Decide which operation to use. Look for key words.

"How many in each group?" tells you to divide.

STEP 3 Use mental math. Memorize your division facts.

Divide the total of 30 into 5 equal groups.

$30 \div 5 = 6$

SOLUTION **There are 6 markers in each group.**

COACHED EXAMPLE

Nick has 40 DVDs in his collection. He put 8 DVDs on each shelf. How many shelves did Nick use to store his DVDs?

THINKING IT THROUGH

Identify the information.

____

____

Look for key words.

____

Which operation should you use?

____

Complete the operation to solve the problem.

_____ ◯ _____ = _____

Nick used _____ shelves to store his DVDs.

Lesson Practice

Choose the correct answer.

1. Jon swam 47 laps in the pool on Saturday. He swam 39 laps on Sunday. How many more laps did Jon swim on Saturday than on Sunday?

A. 8

B. 12

C. 18

D. 86

2. Allison used 85 beads for one necklace. She used 74 beads for another necklace. How many beads did Allison use in all?

A. 149

B. 151

C. 159

D. 160

3. Regina bought 4 packages of erasers. Each package has 2 erasers. How many erasers did Regina buy?

A. 2

B. 4

C. 6

D. 8

4. Dennis picked 18 flowers. He put the flowers in 6 vases. Each vase has the same number of flowers. How many flowers are in each vase?

A. 3

B. 9

C. 12

D. 24

5. Xavier put 3 stickers on each holiday card he made. He made 9 cards. How many stickers did Xavier use?

A. 12

B. 17

C. 27

D. 39

6. Lisa practiced the drums for 135 minutes the first weekend. The next weekend, she practiced for 152 minutes. How many fewer minutes did Lisa practice the drums the first weekend than the second?

A. 17 minutes

B. 27 minutes

C. 187 minutes

D. 287 minutes

7. A plane flew 489 miles from one city to the next. It then turned around and flew back to the first city. How many total miles did the plane fly in the round trip?

Answer ______________ miles

8. Stephen separated 32 books into 4 equal piles. How many books are in each pile?

Answer ______________

EXTENDED-RESPONSE QUESTION

9. Dena runs 4 days each week. She runs 7 miles on each of the 4 days.

Alan runs 3 days each week. He runs 9 miles on each of the 3 days.

Part A How many miles does Dena run each week?
How many miles does Alan run each week?

____

____

Part B Who runs more miles each week? By how many miles more?

____

____

Part C Explain how you solved each part of the problem.

____

____

____

1 Review

1 Which number is the same as five hundred twenty?

A 52

B 520

C 250

D 5,200

2 The model of a number is shown below.

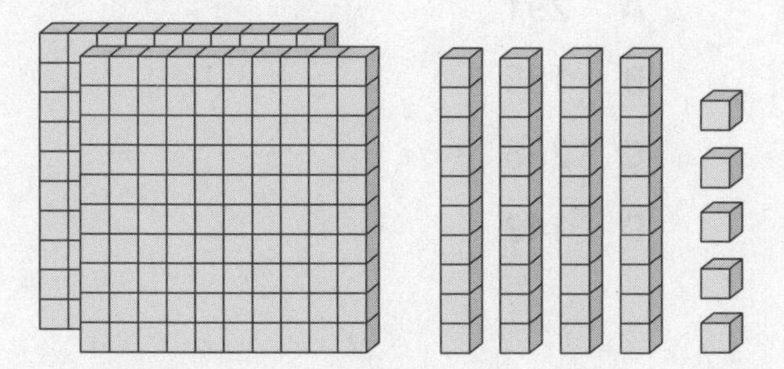

Which number does the model represent?

A 145

B 542

C 254

D 245

3 A team's bowling scores are shown in the table below.

BOWLING SCORES

Player	Score
Lucia	116
Noah	125
Madison	189
Ben	214

Who scored an even number of points?

A Noah and Ben

B Lucia and Ben

C Madison and Noah

D Ben and Noah

4 Rebecca has 18 beads in her bracelet. Half of the beads are blue. How many beads are not blue?

A 17

B 10

C 9

D 8

5 Tyler sees this number sentence on the board.

$7 + 8 + 3 = \_\_\_\_\_ + 7 + 3$

What number belongs on the line to make the number sentence true?

A 3

B 5

C 7

D 8

6 The table shows the number of boxes of cookies sold by each student.

COOKIES SOLD

Student	Number of Boxes
Anne	285
James	303
Stan	319
Charla	297

Who sold the **most** boxes of cookies?

A Anne

B James

C Stan

D Charla

7 Which number belongs in the box to make the number sentence true?

$738 > \square$

A 729

B 738

C 741

D 744

8 To the nearest hundred, about 200 people attended the show. Which could be the number of people who attended the show?

A 291

B 185

C 266

D 142

9 Billy drew a rectangle. He shaded $\frac{3}{6}$ of the rectangle gray.

Which fraction is equivalent to $\frac{3}{6}$?

A $\frac{1}{3}$

B $\frac{1}{2}$

C $\frac{2}{3}$

D $\frac{3}{5}$

10 A train traveled 229 miles from New York City to Washington, DC. It then traveled 195 miles from Washington, DC to Norfolk, VA. About how may miles did the train travel from New York to Virginia?

A 200 miles

B 400 miles

C 500 miles

D 600 miles

11 Maxine writes the number sentence below.

$$2 \times 6 = 6 \times \_\_\_\_$$

What number belongs on the line to make the number sentence true?

A 0

B 2

C 4

D 6

12 Henry put 12 stickers on 3 pages. There is the same number of stickers on each page. Which model **best** shows Henry's stickers?

A 

B 

C 

D

13

$$\begin{array}{r} 524 \\ -\ 186 \\ \hline \end{array}$$

A 710

B 348

C 338

D 462

14 Which rectangle is $\frac{1}{6}$ shaded?

A

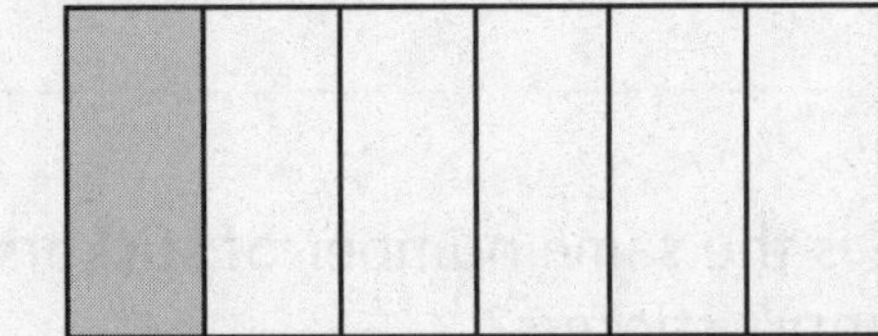

B

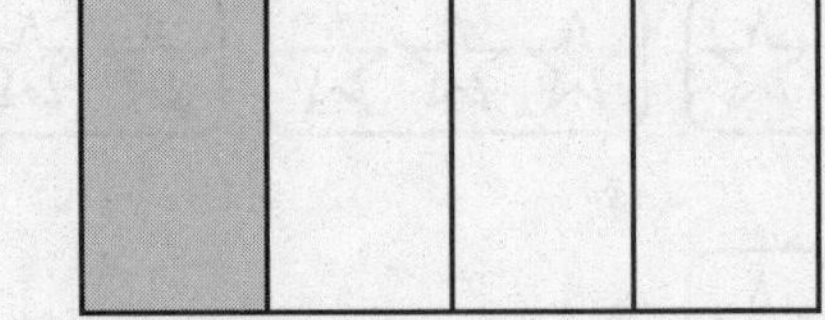

C

D

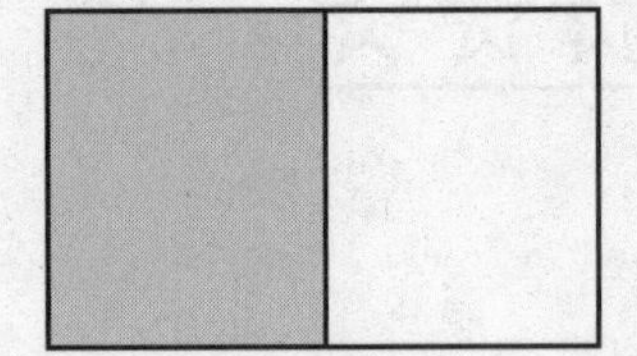

15 Alyssa has 357 stamps in her collection. Anthony has 200 more stamps than Alyssa in his collection. How many stamps are in Anthony's collection?

A 157

B 557

C 467

D 387

16 Which circle is $\frac{5}{8}$ shaded?

A

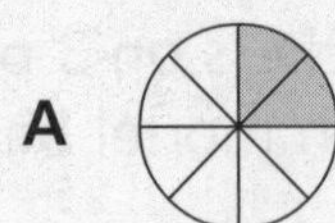

B

C

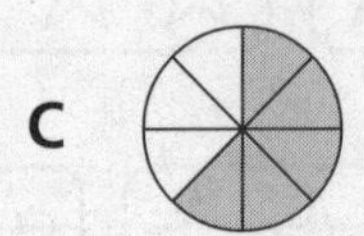

D

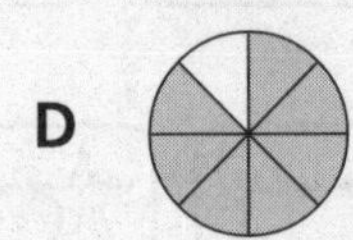

17 Multiply.

$$3 \times 8 = \square$$

A 11

B 12

C 16

D 24

18 Chauncy has the bag of marbles shown below.

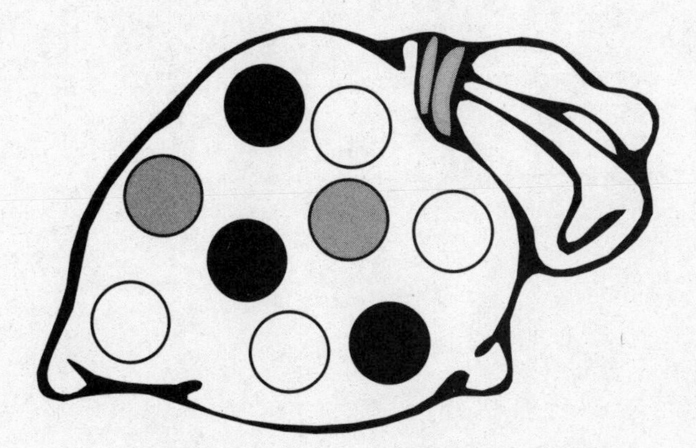

What fraction of the marbles are black?

A $\frac{3}{3}$

B $\frac{2}{3}$

C $\frac{3}{6}$

D $\frac{3}{9}$

19 The students are standing in 6 lines. Each line has 12 students. How many students are standing in line?

A 72 C 36

B 48 D 18

20 What number belongs on the line to make the number sentence true?

$$10 \times \_\_\_\_ = 10$$

A 0 C 11

B 1 D 100

21 Which is the **greatest** fraction?

A $\frac{1}{10}$ C $\frac{1}{2}$

B $\frac{1}{4}$ D $\frac{1}{3}$

22 A sporting goods store sold 237 pairs of sneakers on Saturday and 294 pairs on Sunday. How many pairs of sneakers did the store sell in two days in all?

Show your work.

Answer ____________ pairs of sneakers

23 Ann Marie skip counted these numbers.

150, 175, 200, ☐, 250, 275

What is the missing number?

Show your work.

Answer ______________

24 Ned and his family went on a trip. On the first day, they traveled 236 miles. On the second day, they traveled 282 miles.

Part A

Estimate the number of miles Ned and his family traveled in two days.

Show your work.

Answer ______________ miles

Part B

On the lines below, explain why your estimate in Part A is reasonable.

____

____

____

STRAND

2 Algebra

NYS Math Indicators

Lesson

18 Use Symbols to Compare Fractions

Getting the Idea

Use these symbols when comparing fractions, or any other numbers.

$>$ means **is greater than**.

$<$ means **is less than**.

$=$ means **is equal to**.

EXAMPLE 1

Which symbol belongs in the ○ to make the sentence true? Write $<$, $>$, or $=$.

$\frac{1}{3} \bigcirc \frac{1}{5}$

STRATEGY **Make a model. Choose the correct symbol.**

STEP 1 Make a model to show each fraction.

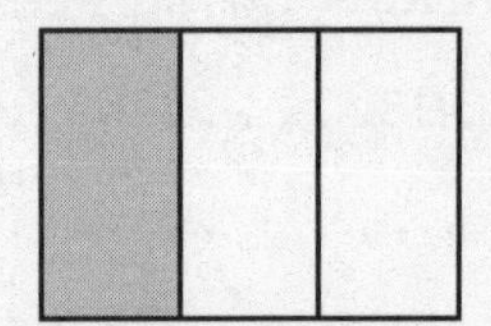

$\frac{1}{3}$

$\frac{1}{5}$

STEP 2 More area is shaded for $\frac{1}{3}$ than for $\frac{1}{5}$.

So $\frac{1}{3}$ is greater than $\frac{1}{5}$.

STEP 3 Choose the correct symbol.

$>$ means is greater than.

SOLUTION $\frac{1}{3} \textcircled{>} \frac{1}{5}$

EXAMPLE 2

Which symbol belongs in the ◯ to make the sentence true? Write $<$, $>$, or $=$.

$\frac{1}{6} \bigcirc \frac{1}{4}$

STRATEGY **Use number lines.**

STEP 1 Make two number lines from 0 to 1.

One number line is in sixths and the other in fourths.

Find $\frac{1}{6}$ and $\frac{1}{4}$ on the number lines.

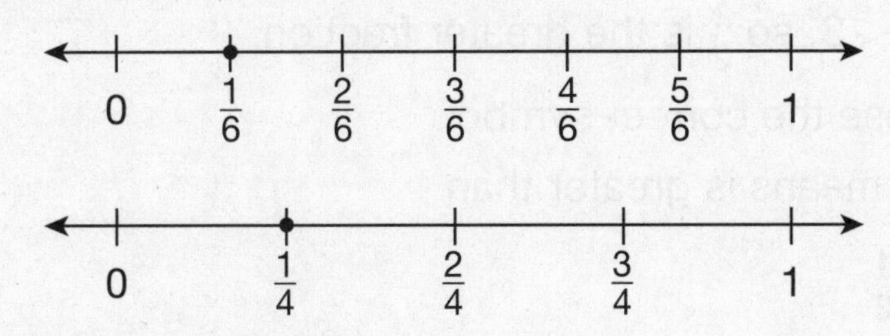

STEP 2 Compare the fractions.

$\frac{1}{6}$ is to the left of $\frac{1}{4}$.

The fraction farther to the left is the lesser fraction.

$\frac{1}{6}$ is the lesser fraction.

STEP 3 Choose the correct symbol.

$<$ means is less than.

SOLUTION $\frac{1}{6} < \frac{1}{4}$

A unit fraction has a 1 in the numerator. When comparing unit fractions, or fractions with the same numerator, just compare the denominators. The fraction with the greater denominator is the lesser fraction.

In Example 2, $\frac{1}{6}$ and $\frac{1}{4}$ have the same numerator.

So compare the denominators: $6 > 4$. $\frac{1}{6} < \frac{1}{4}$

EXAMPLE 3

Which symbol belongs in the ◯ to make the sentence true? Write <, >, or =.

$\frac{1}{2} \bigcirc \frac{1}{3}$

STRATEGY **Compare the fractions.**

STEP 1 Are the numerators the same?

$\frac{1}{2}$ and $\frac{1}{3}$ are both unit fractions. They have 1 in the numerator.

STEP 2 Compare the denominators.

$2 < 3$, so $\frac{1}{2}$ is the greater fraction.

STEP 3 Choose the correct symbol.

$>$ means is greater than.

SOLUTION $\frac{1}{2} \textcircled{>} \frac{1}{3}$

COACHED EXAMPLE

Which symbol belongs in the circle to make the sentence true? Write <, >, or =.

$\frac{1}{5} \bigcirc \frac{1}{8}$

THINKING IT THROUGH

Are the numerators the same? ____________

Both $\frac{1}{5}$ and $\frac{1}{8}$ are unit fractions with _____ as the numerator.

Compare the denominators.

_____ $>$ _____

Is $\frac{1}{5}$ greater than or less than $\frac{1}{8}$? ____________________

Which symbol do you use? _____

$\frac{1}{5}$ _____ $\frac{1}{8}$

Lesson Practice

Choose the correct answer.

1. Which symbol belongs in the ◯ to make the sentence true?

$$\frac{1}{4} \bigcirc \frac{1}{10}$$

A. $<$ **C.** $=$
B. $>$ **D.** $+$

2. Which symbol belongs in the ◯ to make the sentence true?

$$\frac{1}{6} \bigcirc \frac{1}{3}$$

A. $<$ **C.** $=$
B. $>$ **D.** $+$

3. Which symbol belongs in the ◯ to make the sentence true?

$$\frac{1}{2} \bigcirc \frac{1}{4}$$

A. $<$ **C.** $=$
B. $>$ **D.** $+$

4. Which fraction will make this sentence true?

$$\square < \frac{1}{6}$$

A. $\frac{1}{2}$ **C.** $\frac{1}{6}$
B. $\frac{1}{5}$ **D.** $\frac{1}{8}$

5. Which fraction will make this sentence true?

$$\frac{1}{3} < \square$$

A. $\frac{1}{2}$ **C.** $\frac{1}{5}$
B. $\frac{1}{3}$ **D.** $\frac{1}{10}$

6. Which sentence is true?

A. $\frac{1}{4} < \frac{1}{3}$

B. $\frac{1}{6} > \frac{1}{2}$

C. $\frac{1}{8} > \frac{1}{5}$

D. $\frac{1}{10} < \frac{1}{10}$

7. Which symbol belongs in the ◯ to make the sentence true? Write $<$, $>$, or $=$.

$$\frac{1}{5} \bigcirc \frac{1}{6}$$

Answer ______________

8. Which symbol belongs in the ◯ to make the sentence true? Write $<$, $>$, or $=$.

$$\frac{1}{10} \bigcirc \frac{1}{2}$$

Answer ______________

Lesson

19 Patterns with Numbers

Getting the Idea

A **pattern** is a set of numbers or shapes that follows a **rule**. The rule tells you how the numbers or shapes are related.

A **number pattern** uses only numbers.

When the numbers are increasing, use addition.

When the numbers are decreasing, use subtraction.

Look at the pattern below.

20, 25, 30, 35, 40, 45

Rule: Add 5 to the previous number.

EXAMPLE 1

What is the rule of this number pattern?

22, 18, 14, 10, 6, 2

STRATEGY **Find if the numbers increase or decrease and by how much.**

STEP 1 Find if the numbers increase or decrease.

The numbers decrease.

STEP 2 Find how many are between the first two numbers, 22 and 18.

18 is 4 less than 22. Try subtracting 4 from each number.

22 − **4** = 18

18 − **4** = 14

14 − **4** = 10

10 − **4** = 6

6 − **4** = 2

4 is subtracted from each number to get the next number.

STEP 3 Write the rule.

The rule is subtract 4.

SOLUTION **The rule of the pattern is subtract 4.**

You can use a rule to find a missing number or to continue a pattern.

EXAMPLE 2

Julian saw these mailboxes along one side of Poplar Street.

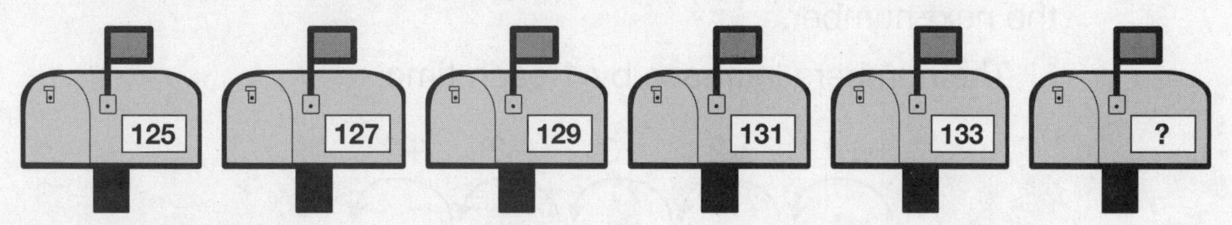

If the pattern continues, what number will be on the next mailbox?

STRATEGY **Find the rule of the pattern. Continue the pattern.**

STEP 1 Do the numbers increase or decrease?

The numbers increase.

STEP 2 Find how many are between the first two numbers, 125 and 127.

127 is 2 more than 125. Try adding 2 to each number.

125 + **2** = 127
127 + **2** = 129
129 + **2** = 131
131 + **2** = 133

The rule is add 2 to each number to get the next number.

STEP 3 Use the rule to continue the pattern.

133 + **2** = 135

SOLUTION **If the pattern continues, the number on the next mailbox will be 135.**

EXAMPLE 3

What is the missing number in this number pattern?

5, 6, 8, 11, ?, 20

STRATEGY **Find the rule of the pattern.**

STEP 1 Do the numbers increase or decrease?

The numbers increase.

STEP 2 Look at the numbers that are added to each number to get the next number.

The numbers increase by 1 each time.

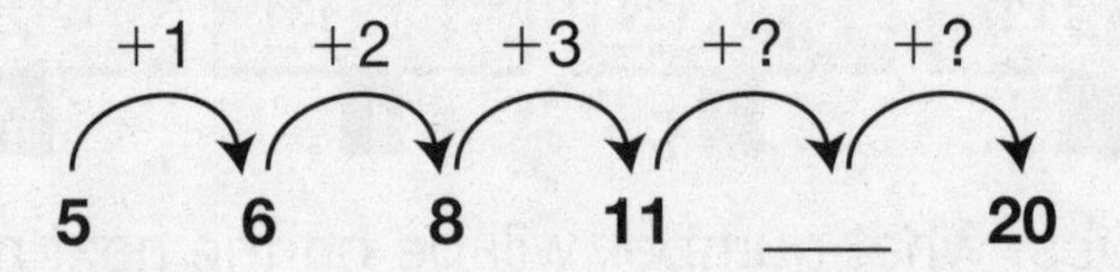

STEP 3 Find the number that is added to 11 to get the missing number.

3 was added to 8, so add 4 to 11.

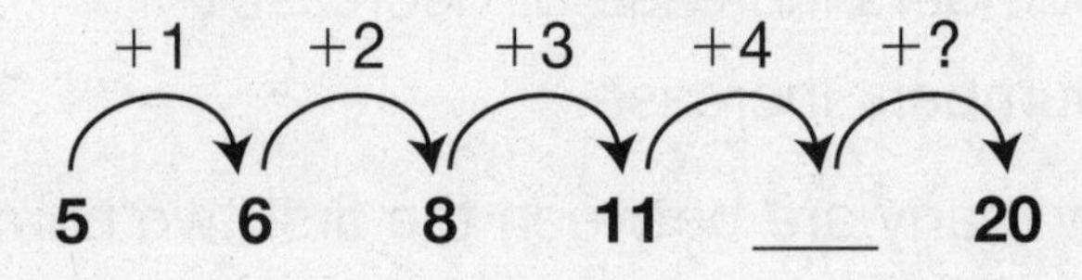

$11 + 4 = 15$

STEP 4 Check to make sure the last number in the pattern is 20.

4 was added to 11, so now add 5 to 15.

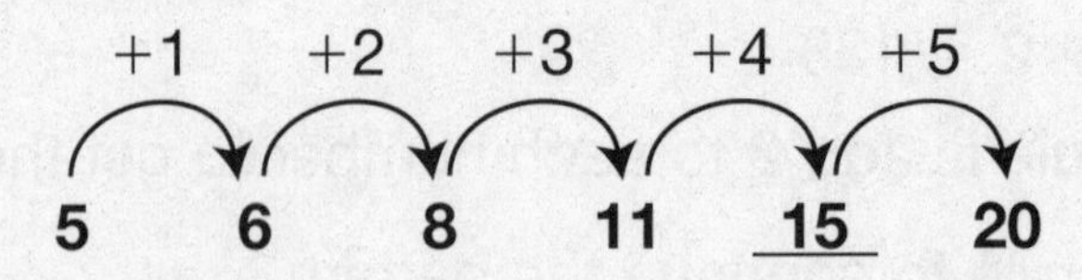

$15 + 5 = 20$

SOLUTION **The missing number is 15.**

COACHED EXAMPLE

What is the next number in this number pattern?

34, 31, 28, 25, 22, __?__

THINKING IT THROUGH

Do the numbers increase or decrease? ____________________

Find how many are between the first two numbers.

31 is ________ less than 34.

Try subtracting ________ from each number.

34 − ________ = ________

31 − ________ = ________

28 − ________ = ________

25 − ________ = ________

The rule is ____________________.

Use the rule to find the next number in the pattern.

22 − ________ = ________

The next number in the pattern is __________.

Lesson Practice

Choose the correct answer.

1. Francine made the following pattern.

54, 47, 40, 33, 26, 19

What is the rule for Francine's number pattern?

A. add 7

B. add 8

C. subtract 7

D. subtract 8

2. What is the next number in this number pattern?

7, 11, 15, 19, 23, ____

A. 24

B. 25

C. 26

D. 27

3. Marcus made a pattern using the rule add 5. Which could be Marcus's number pattern?

A. 5, 10, 14, 17, 19, 20

B. 6, 11, 16, 21, 26, 31

C. 7, 11, 15, 19, 23, 27

D. 8, 13, 18, 24, 30, 35

4. What is the missing number in this pattern?

14, 17, ____, 23, 26, 29

A. 18

B. 20

C. 21

D. 22

5 What is the next number in this pattern?

17, 15, 13, 11, 9, ____

A. 7

B. 8

C. 9

D. 15

6. What is the missing number in this pattern?

____, 4, 8, 14, 22, 32

A. 2

B. 3

C. 4

D. 5

7. What is the next number in this pattern?

1, 7, 13, 19, 25, ____

Answer __________

8. What is the missing number in this pattern?

37, 34, 31, 28, ____, 22

Answer __________

EXTENDED-RESPONSE QUESTION

9. Mohammed began a workout routine. His workout was 15 minutes on the first day, 18 minutes on the second day, 21 minutes on the third day, 24 minutes on the fourth day, and 27 minutes on the fifth day.

Part A The pattern of workout minutes is 15, 18, 21, 24, 27. What is the rule for the pattern?

____

Part B If the pattern continues, how many minutes will the workout be on the sixth day?

____

Part C Explain how you decided what rule to use and how you continued the pattern.

____

____

____

Lesson

Patterns with Shapes

Geometric patterns are made up of shapes or objects arranged in a certain way.

Like a number pattern, a pattern with shapes also follows a rule. The shapes can form a repeating pattern or a growing pattern.

EXAMPLE 1

What is the next figure in this pattern?

STRATEGY **Decide if it is a growing pattern or a repeating pattern.**

STEP 1 Decide if the figures in the pattern repeat or grow.

The figures in the pattern repeat.

STEP 2 Find the figures that repeat.

Each figure is turned $\frac{1}{4}$ turn from the one before it.

So the first 4 figures repeat.

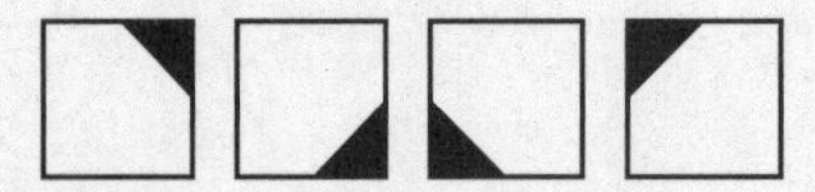

STEP 3 Find the next figure.

The last figure shown is shaded in the top left corner.

So the next figure will be shaded in the top right corner.

SOLUTION **The next figure in the pattern is** **.**

EXAMPLE 2

How many dots will be in the fifth figure of the pattern? Draw the figure.

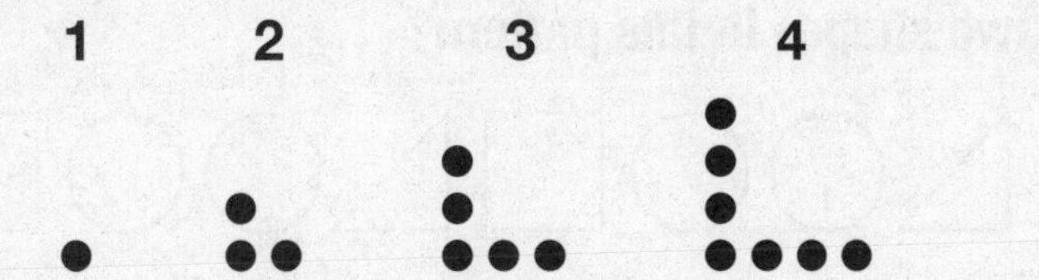

STRATEGY **Use a table to find the rule.**

STEP 1 Decide if the figures in the pattern repeat or grow.

The figures in the pattern grow.

STEP 2 Find the rule of the pattern. Make a table.

Figure	1	2	3	4
Number of Dots	1	3	5	7

The number of dots increases by 2.

One dot goes at the top left. One dot goes at the bottom right.

STEP 3 Use the rule to find how many dots will be in the next figure.

Continue the pattern in the table.

Figure	1	2	3	4	5
Number of Dots	1	3	5	7	9

The figure will have 9 dots.

STEP 4 Draw the next figure.

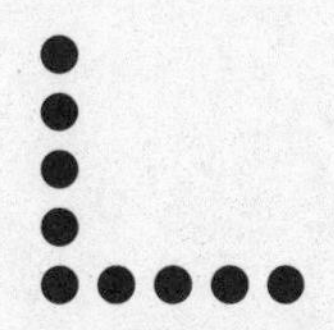

SOLUTION **The fifth figure will have 9 dots. The figure is shown in Step 4.**

COACHED EXAMPLE

What are the next two shapes in this pattern?

○ ○ □ ◺ ○ ○ □ ◺ ○ ○ □ ___?___ ___?___

THINKING IT THROUGH

Find the rule.

Do the shapes in the pattern repeat or grow? ________________

What are the repeating shapes? ______________________________________

____

Use the rule to find the next 2 shapes.

After the square comes a ________________.

After the triangle comes a ________________.

The next two shapes in this pattern are ________________ and ________________.

Lesson Practice

Choose the correct answer.

1. What is the next figure in this pattern?

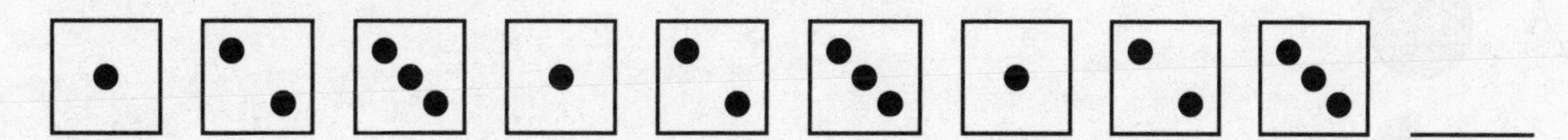

A.

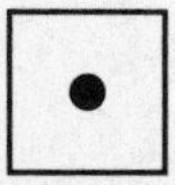

B.

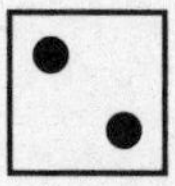

C.

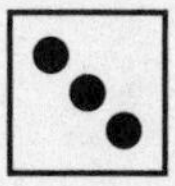

D.

2. What is the missing figure in this pattern?

↑ ↓ ← __ ↑ ↓ ← → ↑ ↓ ← →

A. ←

B. →

C. ↑

D. ↓

3. What is the missing figure in this pattern?

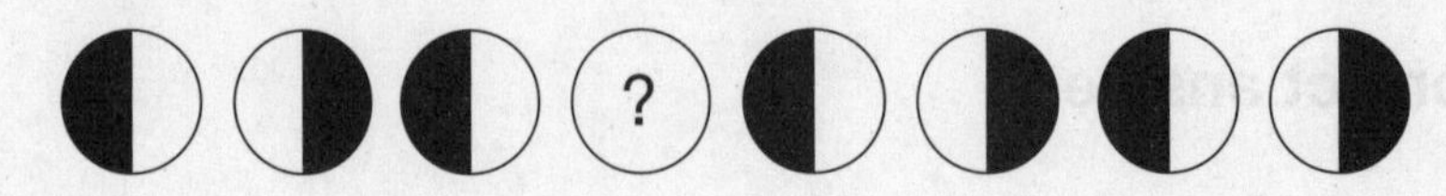

A.

B.

C.

D.

4. Look at this pattern.

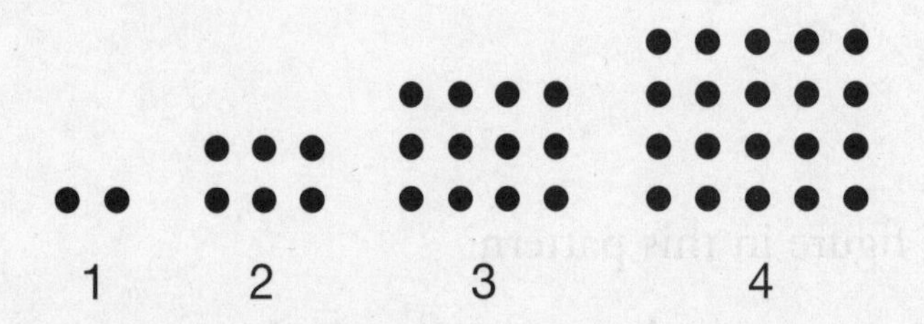

Which is the next figure in the pattern?

A. 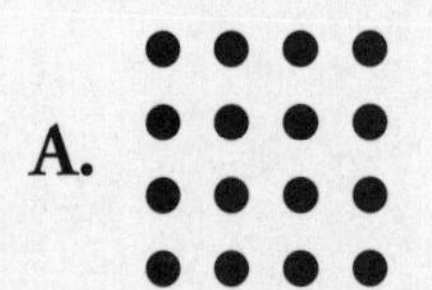

C. 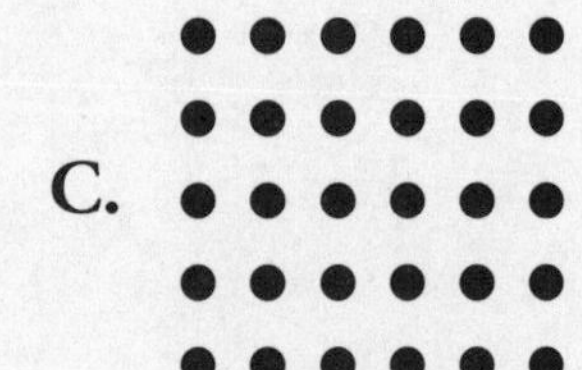

B. 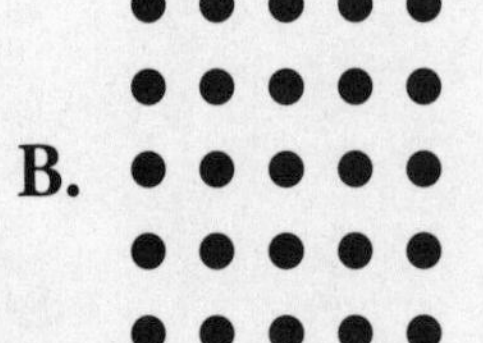

D. 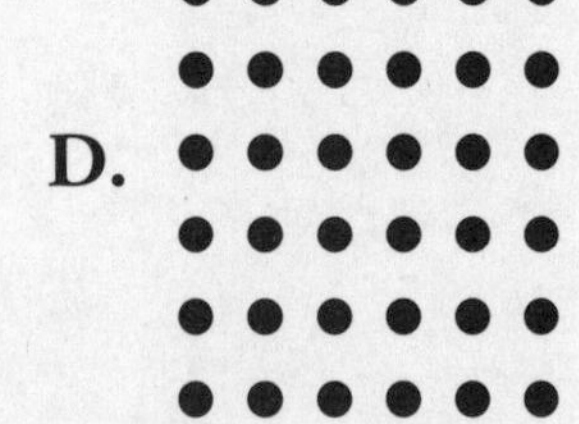

5. How many circles will be in the next figure of this pattern?

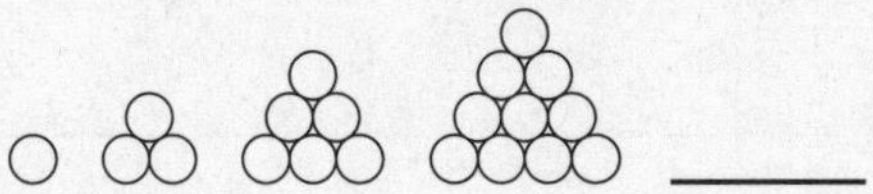

______

Answer ______

EXTENDED-RESPONSE QUESTION

6. Lars drew the figures in the pattern below.

☆ △ ☆ ○ ☆ △ ☆ ○ ☆ ?___ ☆ ○

Part A Draw the missing shape in this pattern.

Part B Explain how you decided which figure to draw.

______

______

______

2 Review

1 Cory wrote a number pattern.

12, 15, 18, 21, 24, 27

Which is the rule to Cory's number pattern?

A Add 3.

B Add 2.

C Subtract 3.

D Subtract 2.

2 What is the missing number in this number pattern?

23, 25, 27, 29, __?__, 33

A 32

B 31

C 34

D 30

3 Which symbol makes this sentence true?

$\frac{1}{6}$ ____ $\frac{1}{2}$

A $>$

B $<$

C $=$

D $\times$

4 Which number pattern uses the rule "add 7"?

A 9, 15, 21, 27, 33, 39

B 7, 14, 22, 29, 37, 42,

C 8, 15, 22, 29, 36, 43

D 7, 15, 22, 30, 37, 43

5 What is the missing number in this number pattern?

32, 28, 24, __?__, 16, 12

A 8

B 20

C 18

D 28

6 Brian made a number pattern starting with 3. The number he adds to get to the next one increases by 2 each time. Which number pattern could be Brian's pattern?

A 3, 6, 12, 24, 48, 96

B 3, 6, 9, 12, 15, 18

C 3, 4, 6, 9, 13, 18

D 3, 5, 9, 15, 23, 33

7 The table shows the number of free throws Lance made in basketball games each year.

LANCE'S FREE THROWS

Year	Free Throws Made
1	20
2	26
3	32
4	38
5	44

If the pattern continues, which shows the number of free throws Lance will probably make in year 6?

A 50

B 42

C 46

D 48

8 Leah wrote this pattern on her paper.

70, 75, 80, 85, 90, 95

Which pattern uses the same rule as Leah's pattern?

A 30, 35, 45, 50, 60, 65

B 18, 23, 27, 32, 36, 41

C 24, 29, 34, 39, 44, 49

D 35, 45, 50, 60, 65, 75

9 Ruby created this number pattern.

3, 10, 17, 24, 31, __?__, __?__

Which shows the next two numbers in her pattern?

A 32, 33

B 38, 40

C 38, 46

D 38, 45

10 Which number pattern uses the rule "add 5"?

A 10, 15, 25, 40, 60, 85

B 15, 25, 35, 45, 55, 65

C 12, 17, 22, 27, 32, 37

D 20, 25, 30, 50, 90, 6

11 Lou Ann is making this pattern along the wall.

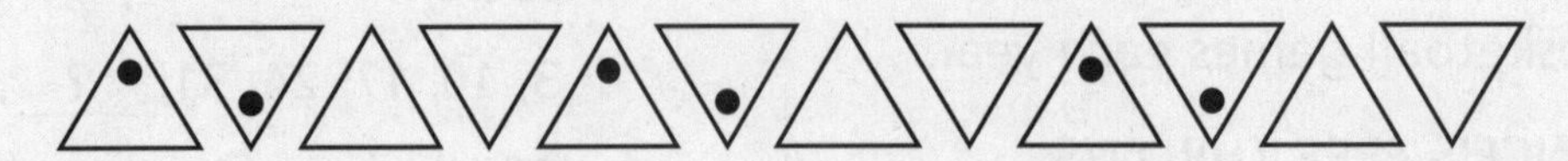

Which shows the next figure Lou Ann will make to continue the pattern?

A ▽ (with dot)

B △ (with dot)

C ▽

D △

12 Mrs. Hall drew this pattern of shapes on the board. The students will continue the pattern.

Which shows the next two shapes in the pattern?

A □△

B △□

C □□

D △△

13 Tristan drew this pattern along the border of his folder. He missed one shape.

△ □ ○ ○ △ □ ? ○ △ □ ○ ○

Which shows the missing shape in the pattern?

A △

B □

C ◇

D ○

14 Alexi started a pattern with these shapes. The first shape in the pattern is the triangle.

Which **best** describes the rule of Alexi's pattern?

A The number of sides decreases by two.

B The number of sides increases by two.

C The number of sides decreases by one.

D The number of sides increases by one.

15 Look at the pattern made with dots.

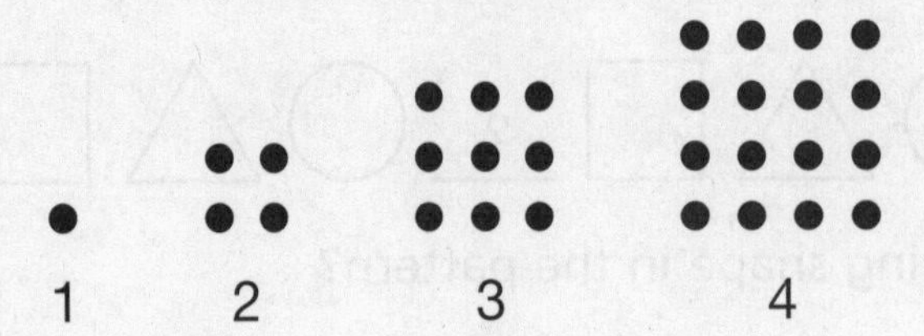

How many dots will be in the **fifth** figure?

A 20

B 25

C 30

D 36

16 Nathan wrote this number pattern.

8, 11, 14, 17, 20, ___?___

What is the next number in the pattern?

Show your work.

Answer ____________________

17 Draw the two missing figures in this pattern.

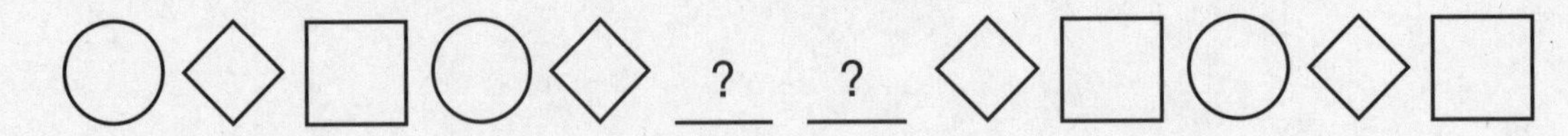

Answer ____________________

18 Yoko and Shawn each made a pattern.

Part A

Yoko made this pattern with numbers.

1, 2, 4, 8, 16, __?__

What is the next number in Yoko's pattern?

Answer ____________________

Part B

Shawn drew this pattern with figures.

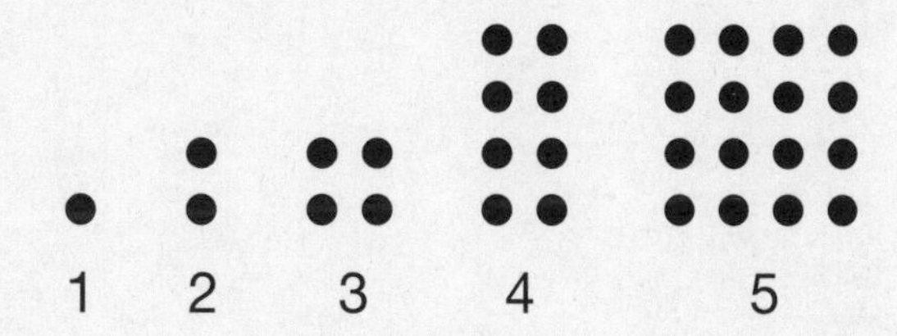

On the space below, draw the next figure to continue Shawn's pattern.

Part C

On the lines below, explain how you decided how many dots to draw in the figure.

____

____

____

STRAND

3 Geometry

21 Two-Dimensional Shapes

Getting the Idea

A **two-dimensional shape** is a flat shape that has length and width.

A **circle** is a two-dimensional shape with all points the same distance from a point called the center. It has no corners and no edges.

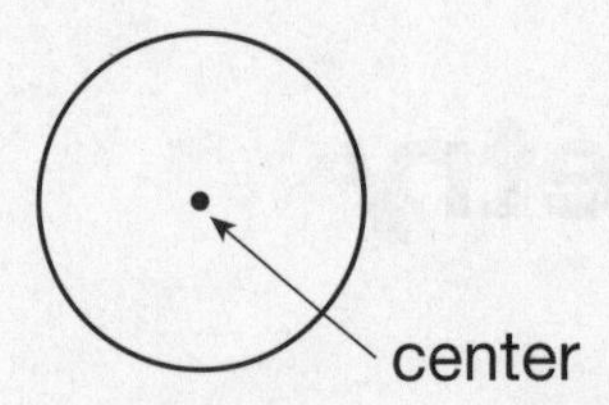

The shapes below have straight sides. They can be named by the number of **sides** and **angles** they have.

A **triangle** is a two-dimensional shape with 3 sides and 3 angles.

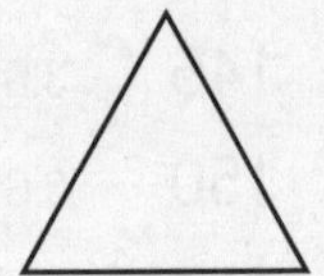 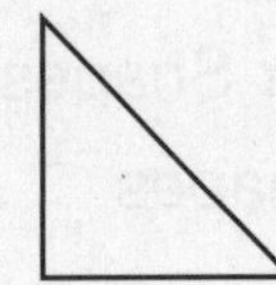 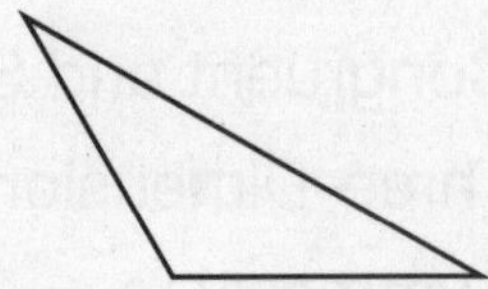

A **rectangle** is a two-dimensional shape with 4 sides and 4 angles.
The opposite sides are the same length.
All the angles form square corners.

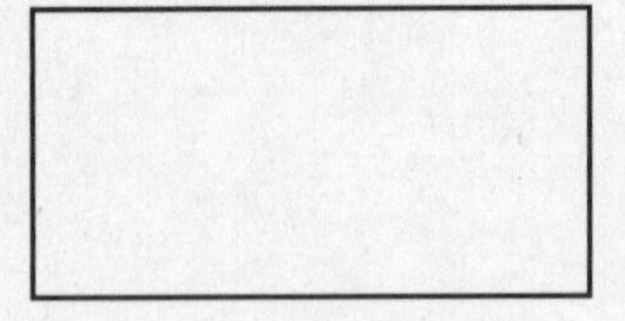 

A **square** is a two-dimensional shape with 4 sides and 4 angles. All 4 sides are the same length. All the angles form square corners.

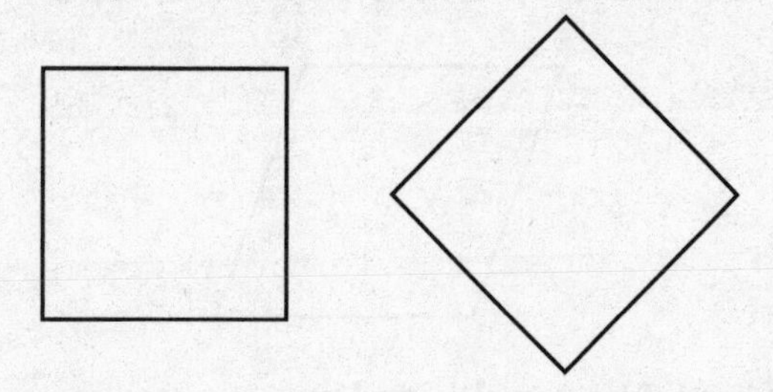

A **trapezoid** is a two-dimensional shape with 4 sides. Not all the angles form square corners.

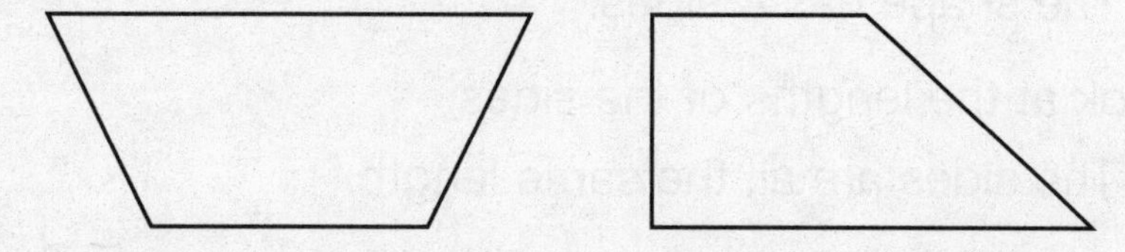

A **rhombus** is a two-dimensional shape with 4 sides the same length.

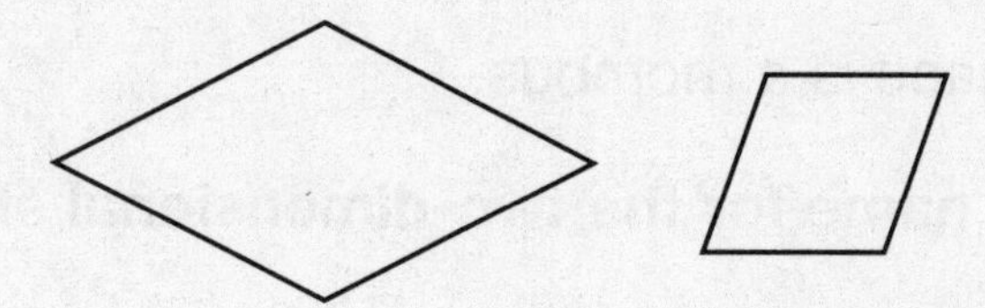

A **hexagon** is a two-dimensional shape with 6 sides and 6 angles.

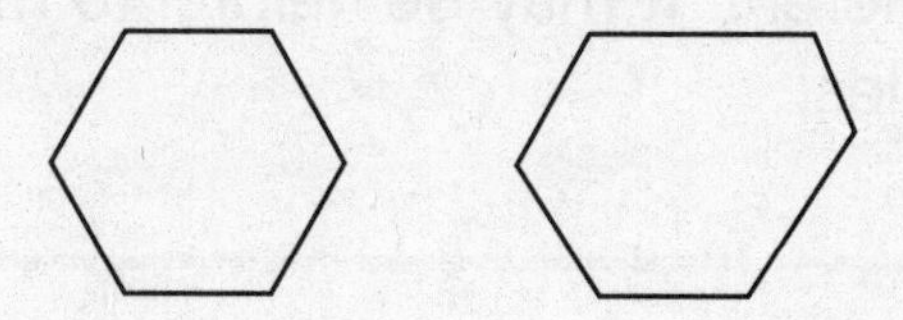

EXAMPLE 1

What is the best name for this two-dimensional shape?

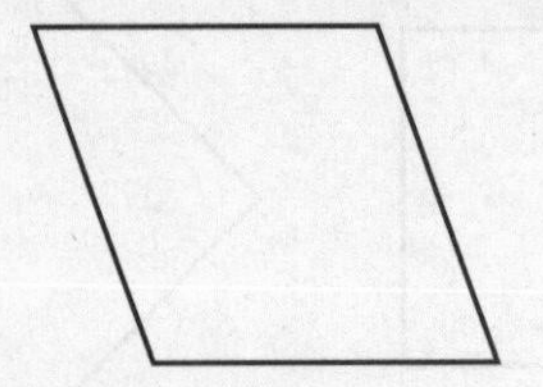

STRATEGY **Look at the sides and angles.**

STEP 1 Count the number of sides.

The shape has 4 sides.

STEP 2 Look at the lengths of the sides.

The sides are all the same length.

STEP 3 Look at the angles.

The angles are not square corners.

The shape is a rhombus.

SOLUTION **The best name for the two-dimensional shape is a rhombus.**

You may be asked to compare two different shapes to understand what makes them alike and different. It may be helpful to make notes about the number of sides and angles.

EXAMPLE 2

List two ways the shapes are the same. List one way the shapes are different. Then name the shapes.

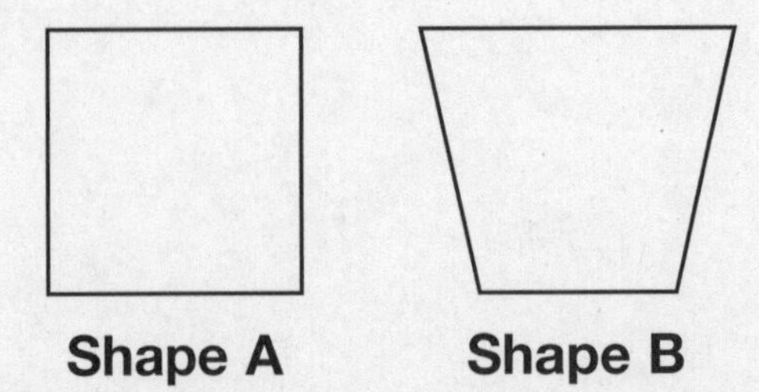

Shape A **Shape B**

STRATEGY **Compare the sides and angles of the shapes.**

STEP 1 Count the number of sides and angles.

Shape A has 4 sides and 4 angles.

Shape B also has 4 sides and 4 angles.

STEP 2 Describe the angles.

Shape A has 4 square corners.

Shape B does not have 4 square corners.

STEP 3 Name the shapes.

Shape A is a square.

Shape B is a trapezoid.

SOLUTION **The shapes are alike because both have 4 sides and 4 angles. The shapes are different because Shape A has 4 square corners and Shape B does not. Shape A is a square. Shape B is a trapezoid.**

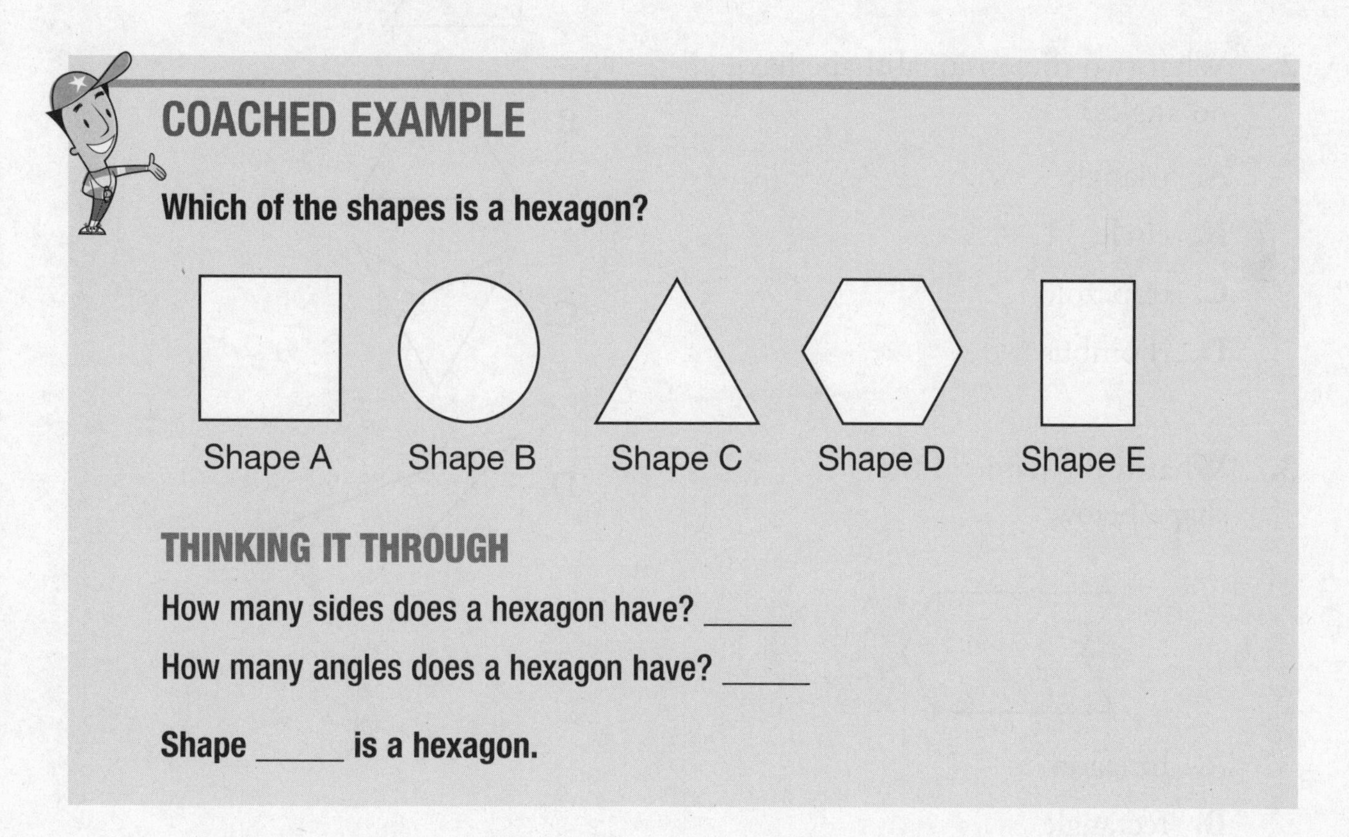

COACHED EXAMPLE

Which of the shapes is a hexagon?

Shape A Shape B Shape C Shape D Shape E

THINKING IT THROUGH

How many sides does a hexagon have? _____

How many angles does a hexagon have? _____

Shape _____ is a hexagon.

Lesson Practice

Choose the correct answer.

1. What is the name of the shape below?

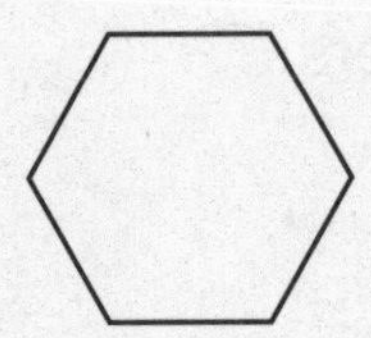

A. hexagon
B. rectangle
C. square
D. triangle

2. What two-dimensional shape has no angles?

A. triangle
B. circle
C. trapezoid
D. rhombus

3. What is the name of the shape below?

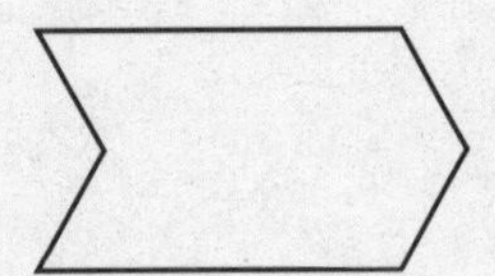

A. hexagon
B. rectangle
C. square
D. triangle

4. Which shape has 1 less side than a rhombus?

A. triangle
B. rectangle
C. hexagon
D. circle

5. Which shape is **not** a triangle?

A. 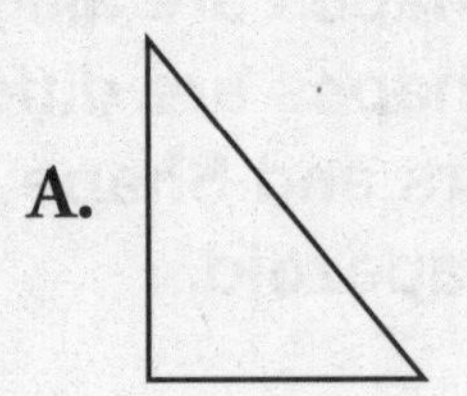

B.

C.

D.

6. Carlos drew a shape that has more angles than a triangle, but fewer angles than a hexagon. Which shape did Carlos draw?

A. 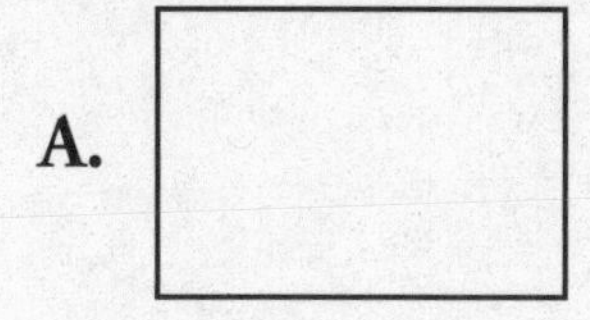

B. 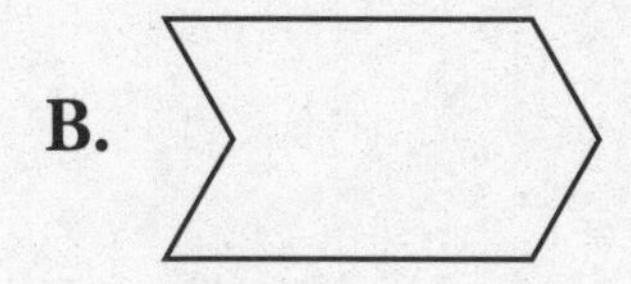

C. 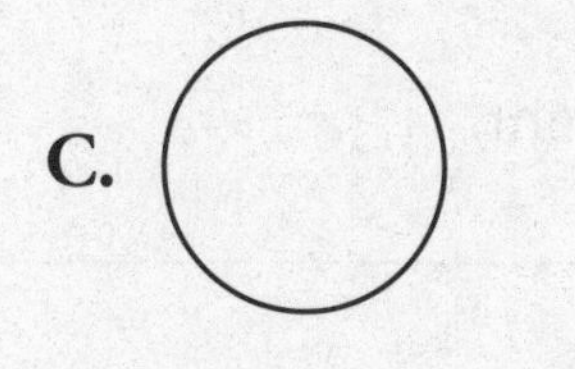

D. 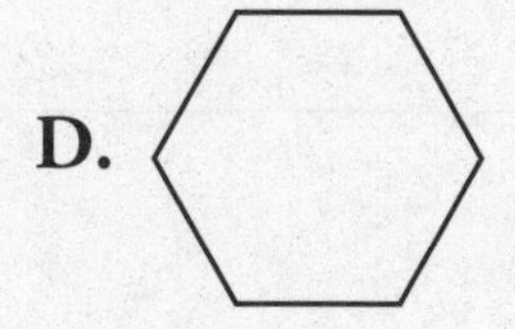

7. How many square corners does a circle have?

A. 0

B. 1

C. 2

D. 3

8. The shape of the window in Richie's bathroom is shown below.

There are a few possible ways to name the shape of the window.

Name the shape in 2 ways.

Answer ____________________

____

9. What is the shape of this yield sign?

Answer ____________________

EXTENDED-RESPONSE QUESTION

10. Nikki saw a triangle and a square.

Part A Draw the two shapes that Nikki saw. Label the shapes.

Part B Explain how the two shapes are alike and different.

___

___

___

Lesson

22 Congruent and Similar Shapes

3.G.2

Getting the Idea

Congruent figures have the same shape and the same size.

Similar figures have the same shape, but can have different sizes.

congruent	similiar not congruent	neither similiar nor congruent

Since congruent figures have the same shape, congruent figures are also similar.

EXAMPLE 1

Are these triangles congruent?

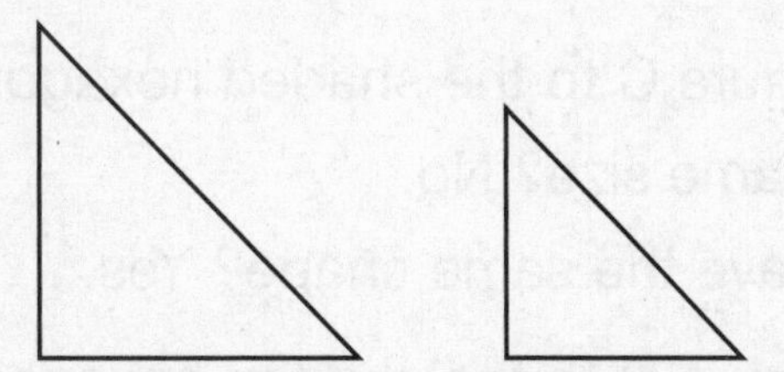

STRATEGY **Decide if the triangles have the same shape and size.**

STEP 1 Do the triangles have the same shape?

Yes, both triangles have the same shape.

STEP 2 Do the triangles have the same size?

No, one triangle is bigger than the other.

SOLUTION **These triangles are not congruent. They are similar.**

EXAMPLE 2

Which figure is congruent to this shaded hexagon?

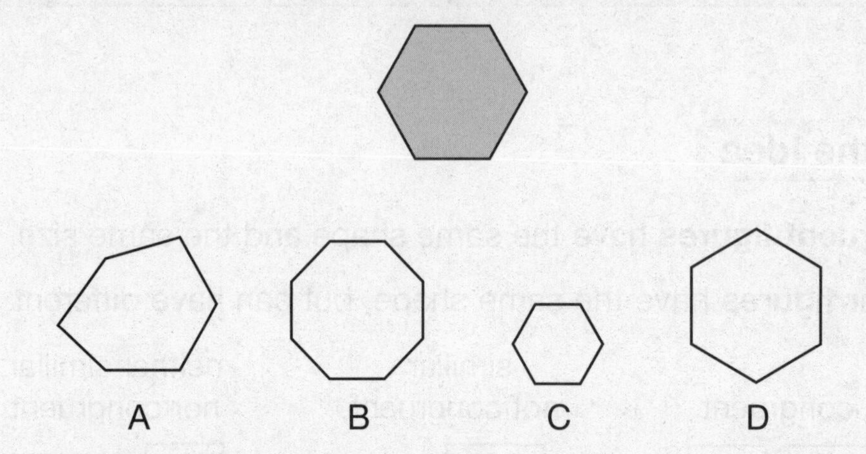

STRATEGY **Compare each figure to the shaded hexagon. Look for a figure with the same size and the same shape.**

STEP 1 Compare Figure A to the shaded hexagon.

Is it the same size? Yes.

Does it have the same shape? No.

STEP 2 Compare Figure B to the shaded hexagon.

Is it the same size? Yes.

Does it have the same shape? No.

STEP 3 Compare Figure C to the shaded hexagon.

Is it the same size? No.

Does it have the same shape? Yes.

STEP 4 Compare Figure D to the shaded hexagon.

Is it the same size? Yes.

Does it have the same shape? Yes.

SOLUTION **Figure D is congruent to the shaded hexagon.**

COACHED EXAMPLE

Are the trapezoids congruent, similar, or neither similar nor congruent?

THINKING IT THROUGH

Do the trapezoids have the same shape? ________________

Do the trapezoids have the same size? ________________

The two trapezoids are ____.

Lesson Practice

Choose the correct answer.

1. Which of the following shows a pair of congruent figures?

A. 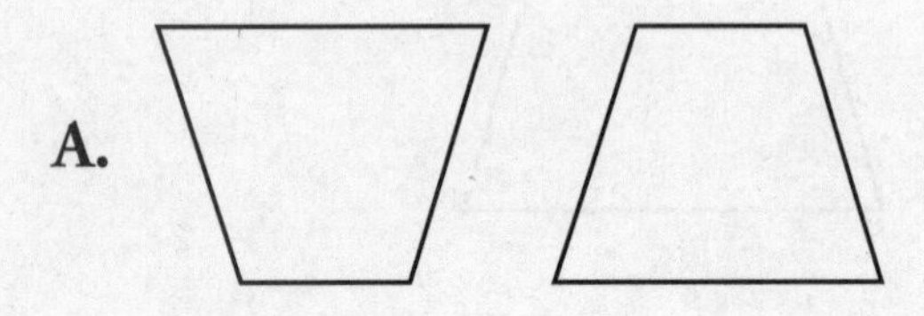

B. 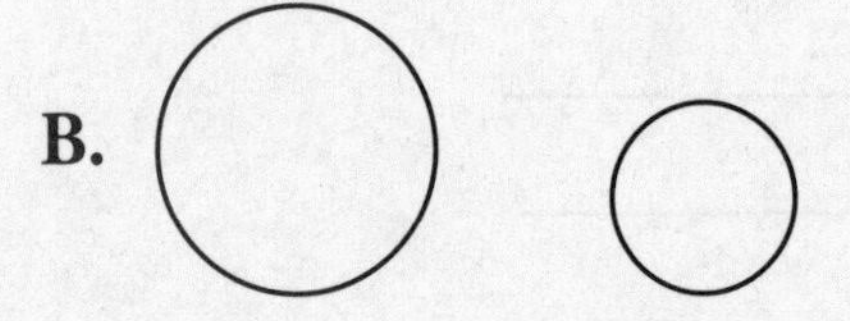

C. 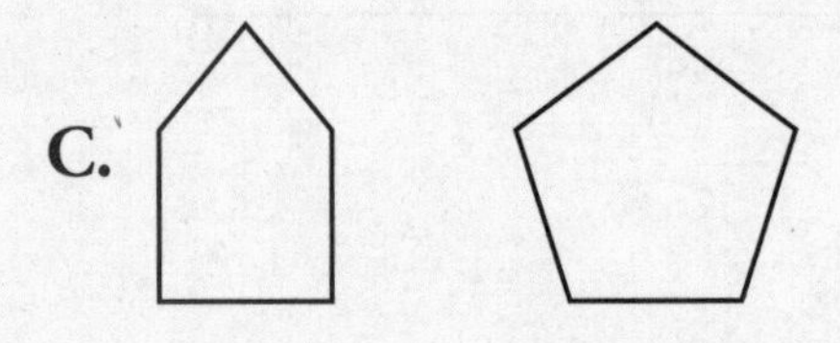

D. 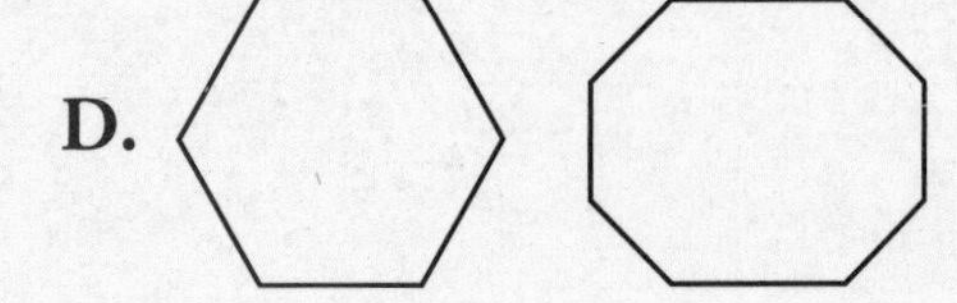

2. Which figure is **not** congruent to this shaded figure?

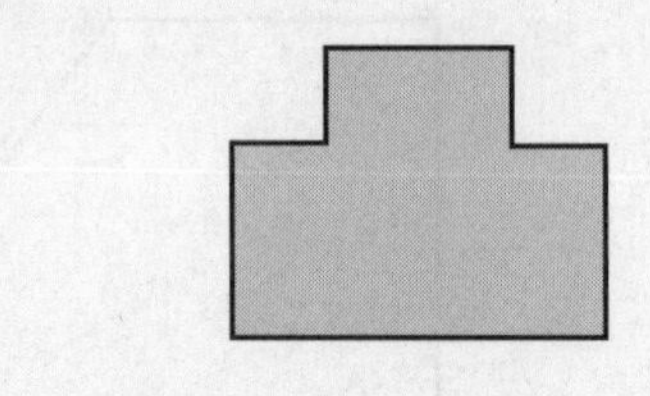

A. 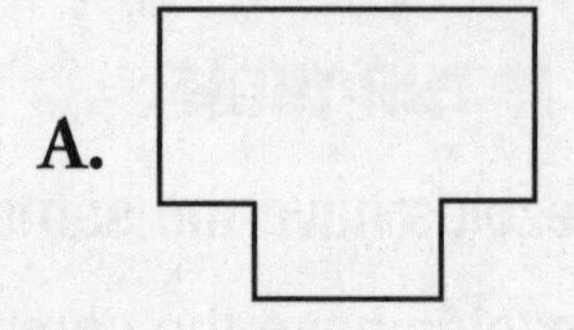

B. 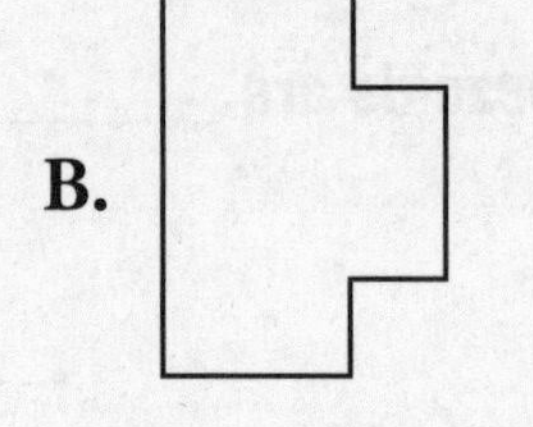

C. 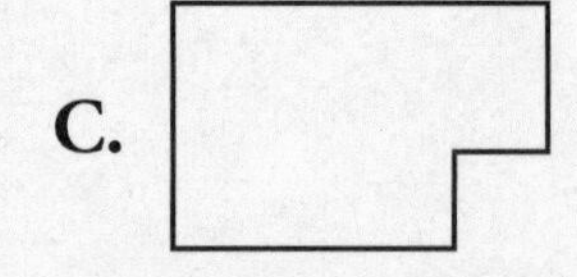

D.

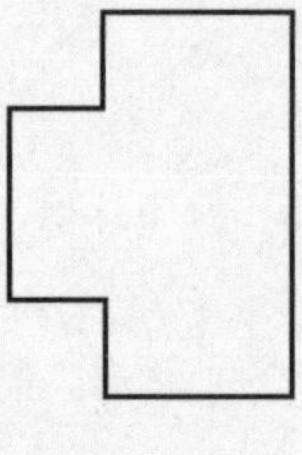

3. Which figure is similar but **not** congruent to the shaded square?

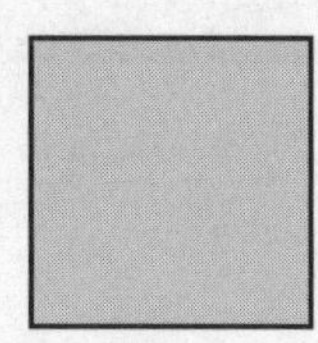

A.

B.

C.

D.

4. Which describes two congruent figures?

A. They have the same size, but not the same shape.

B. They have the same shape, but not the same size.

C. They have the same size and shape.

D. They do not have the same size or shape.

5. Do these figures appear congruent, similar, or neither congruent nor similar?

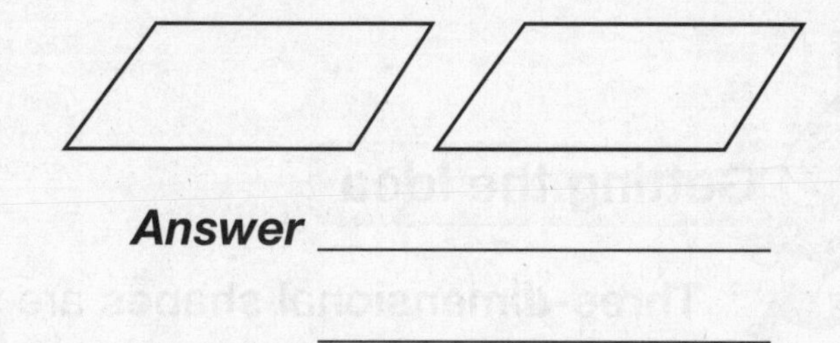

Answer ____________________

____

6. Do these figures appear congruent, similar, or neither congruent nor similar?

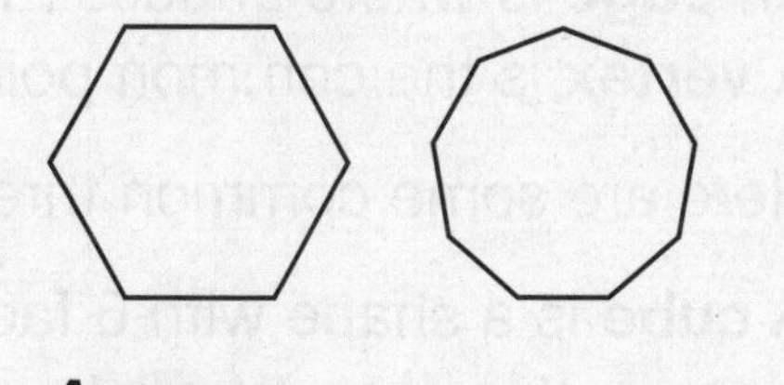

Answer ____________________

____

Lesson

23 Three-Dimensional Shapes

Getting the Idea

Three-dimensional shapes are shapes that have length, width, and height. Everyday objects such as boxes, balls, and cans are three-dimensional shapes. They are also called **solid figures.**

Here are some of the parts in three-dimensional shapes.

A **face** is a flat side of a solid figure.
An **edge** is where 2 faces meet.
A **vertex** is the common point where 3 or more edges meet.

Here are some common three-dimensional shapes.

A **cube** is a shape with 6 faces that are squares. It has 12 edges that are the same length and 8 vertices.

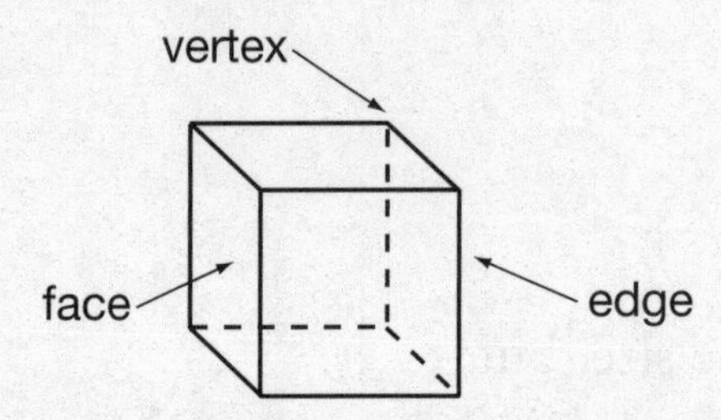

A **rectangular prism** is a shape with 6 faces that are rectangles. It has 12 edges and 8 vertices.

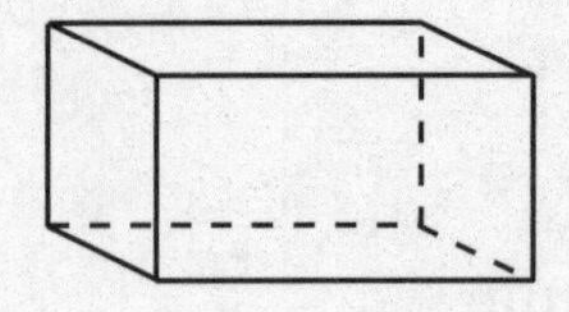

A **triangular prism** is a shape with 5 faces. Two of the faces are triangles and 3 are rectangles. It has 9 edges and 6 vertices.

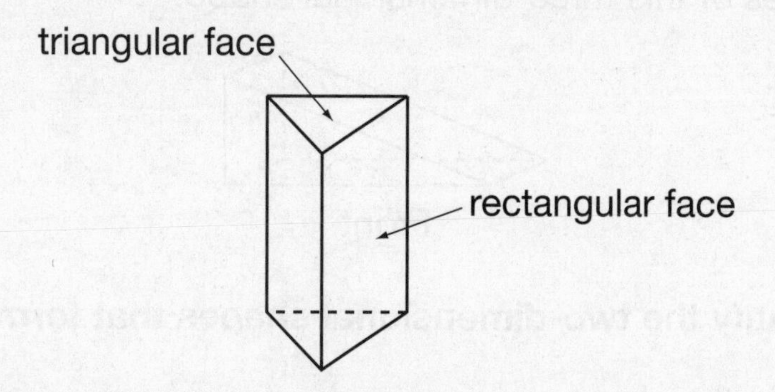

A **sphere** is a round shape with a curved surface. It has no edges and no vertices.

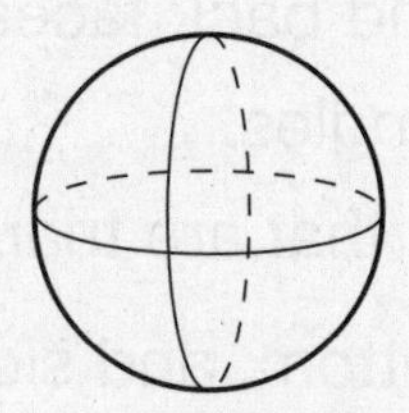

A **cone** is a shape with a curved surface. It has exactly 1 face that is a circle. A cone also has a point opposite the face called a **vertex**.

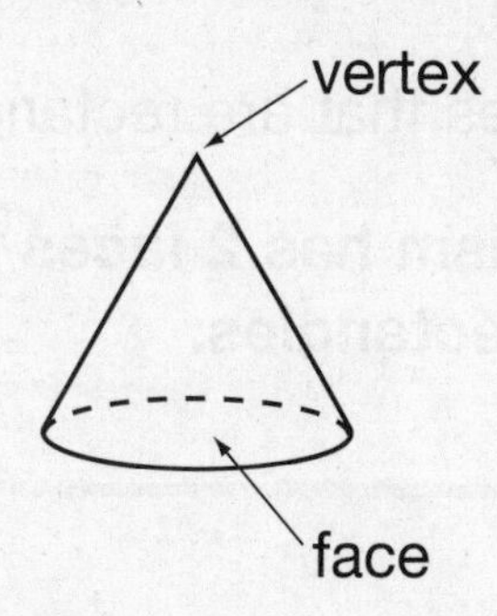

A **cylinder** is a shape with a curved surface. It has 2 faces that are circles.

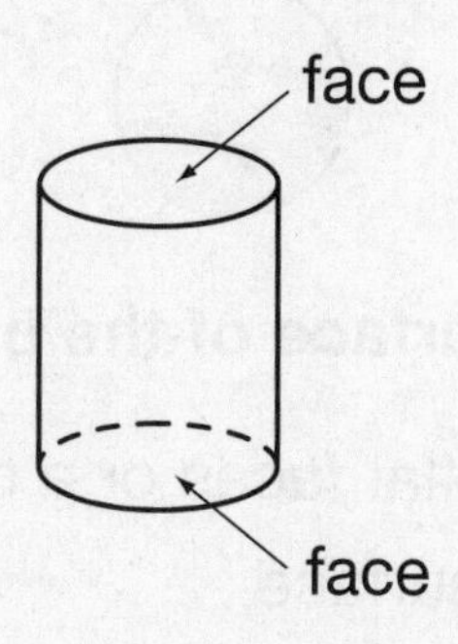

EXAMPLE 1

Describe the faces of this three-dimensional shape.

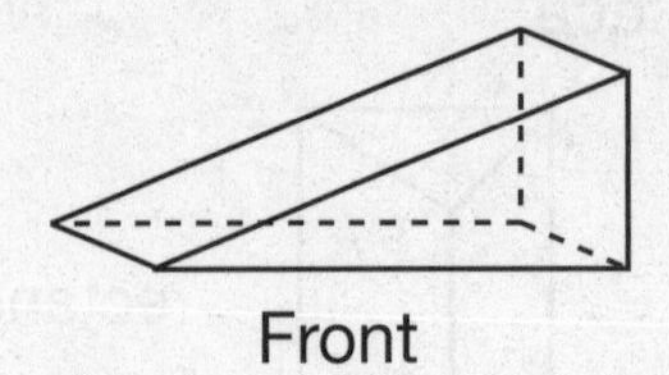

STRATEGY **Identify the two-dimensional shapes that form the faces.**

STEP 1 Name the shape.

The shape is a triangular prism.

STEP 2 Describe the front and back faces.

The faces are triangles.

There are 2 faces that are triangles.

STEP 3 Describe the top, bottom, and side faces.

The top and bottom faces are rectangles.

The side face appears to be a square. A square is also a rectangle.

There are 3 faces that are rectangles.

SOLUTION **This triangular prism has 2 faces that are triangles and 3 faces that are rectangles.**

EXAMPLE 2

What three-dimensional shape is like the shape of this baseball?

STRATEGY **Think about the surface of the baseball.**

STEP 1 Does the ball have flat faces or a curved surface?

It has a curved surface.

STEP 2 Does the baseball have edges or vertices?

It has no edges or vertices.

A sphere is a round shape with no edges and no vertices.

SOLUTION **The baseball is shaped like a sphere.**

EXAMPLE 3

How are these two shapes alike? How are they different?

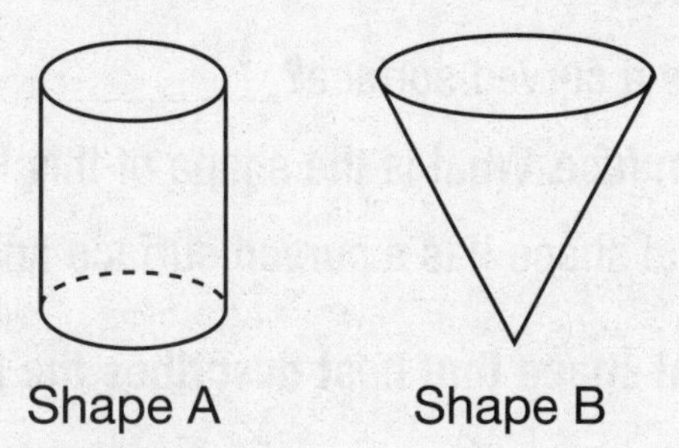

STRATEGY **Compare the surfaces and faces of the shapes.**

STEP 1 Decide if the figures have curved surfaces.

Shape A has a curved surface.

Shape B has a curved surface.

STEP 2 Count the faces and identify the shape of the faces.

Shape A has 2 faces that are circles.

Shape B has 1 face that is a circle.

STEP 3 Name the two shapes.

Shape A is a cylinder.

Shape B is a cone.

SOLUTION **Both are three-dimensional shapes with a curved surface.**
Shape A is a cylinder with 2 faces that are circles.
Shape B is a cone with 1 face that is a circle.

COACHED EXAMPLE

What three-dimensional shape best describes this bullhorn?

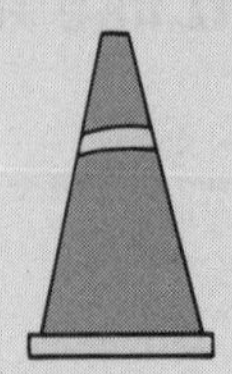

THINKING IT THROUGH

Does the bullhorn have a curved surface? ________

Think about the bottom face. What is the shape of this face? ________________

What three-dimensional shape has a curved surface and 1 flat face? ___________

The three-dimensional shape that best describes the bullhorn is a ______________________.

Lesson Practice

Choose the correct answer.

Use this shape for questions 1 and 2.

1. What is the shape of the soup can?

   A. cone
   B. cube
   C. cylinder
   D. sphere

2. What are the shapes of the faces of the soup can?

   A. squares
   B. circles
   C. triangles
   D. rectangles

3. Which shape does **not** have rectangles as faces?

   A. rectangular prism
   B. triangular prism
   C. cube
   D. cone

4. Which shape has the greatest number of vertices?

   A. 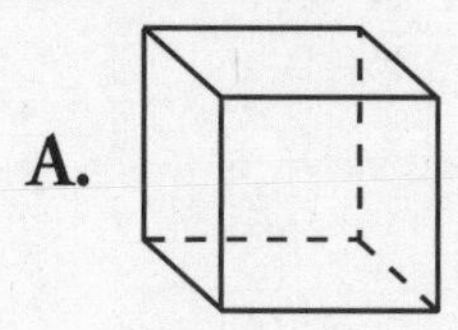

   B. 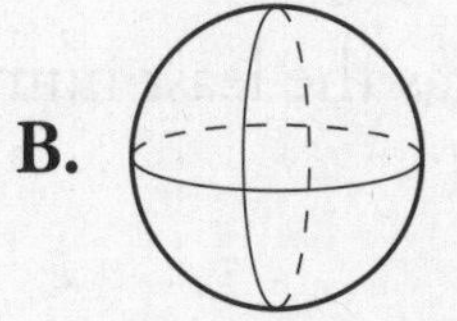

   C. 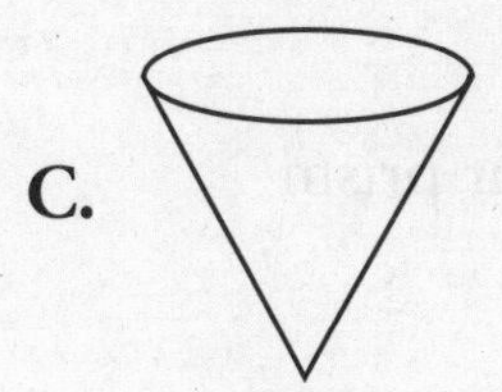

   D. 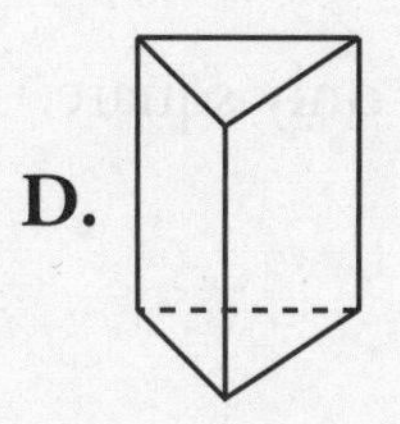

5. Drew picked an ice cube out of his drink. Which of the following best describes a cube?

   A. It is a flat shape.
   B. It has no faces.
   C. All of the faces are the same size.
   D. All the faces are circles.

6. Which shape is least likely to roll off a table?

   A. rectangular prism
   B. sphere
   C. cone
   D. cylinder

7. Which shape has the least number of faces?

   A. cone
   B. cylinder
   C. rectangular prism
   D. cube

8. Which shape has only square faces?

   A. cube
   B. cone
   C. cylinder
   D. sphere

9. Which shape does **not** have vertices?

   A. cube
   B. cone
   C. cylinder
   D. sphere

10. What is the name of this shape?

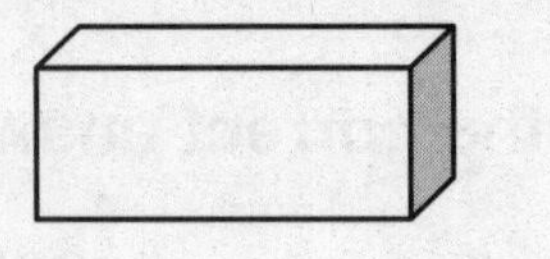

Answer ____________________

11. Camille is describing the shape of a container to a friend.

It has a curved surface.

It has 2 faces.

The faces are circles.

What could be the shape of the container Camille is describing?

Answer ____________________

Lesson

24 Symmetry

Getting the Idea

A shape has **symmetry** if it can be folded on a line so that both sides match exactly. The line is called a **line of symmetry**. Not all shapes have symmetry.

Shapes can have 0, 1, or more than 1 line of symmetry.

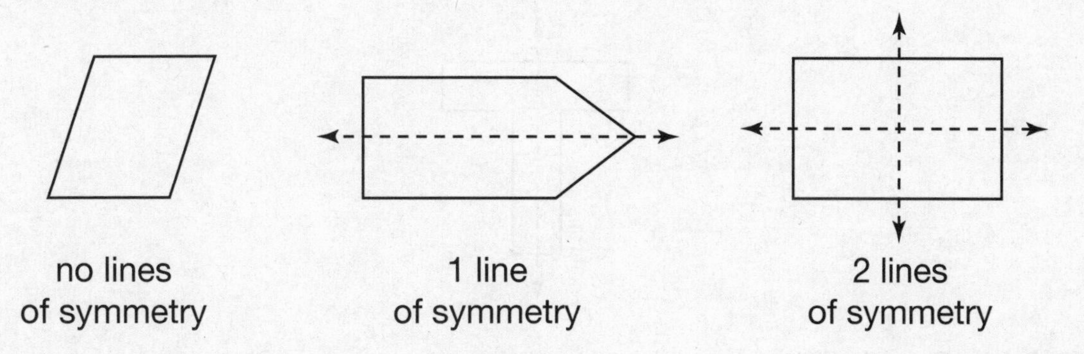

no lines of symmetry | 1 line of symmetry | 2 lines of symmetry

EXAMPLE 1

Is the dotted line a line of symmetry?

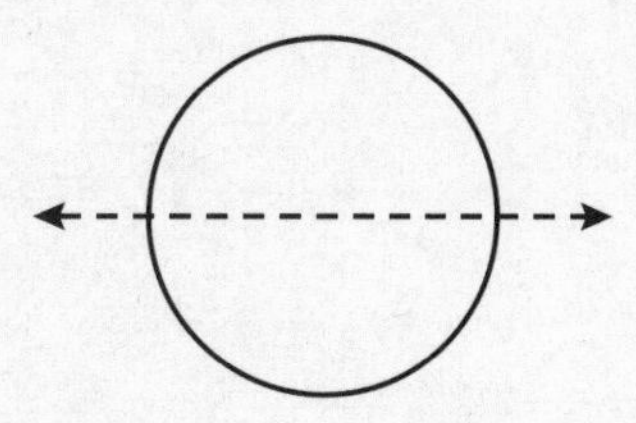

STRATEGY **Imagine folding the circle on the line. See if the sides match exactly.**

Each side is a half circle. The sides match.

The circle has symmetry.

SOLUTION **The dotted line is a line of symmetry.**

EXAMPLE 2

Does this block letter have symmetry? If so, draw a line of symmetry.

STRATEGY **Decide if the letter can be folded on a line and have both sides match exactly.**

STEP 1 Draw a vertical line through the middle of the letter.

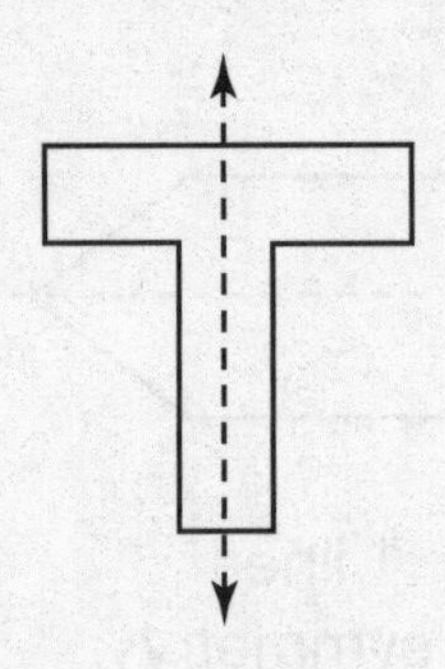

STEP 2 Decide if the two sides match.

The two sides match exactly when the letter is folded.

SOLUTION **This block letter has symmetry. The line of symmetry is shown in Step 1.**

EXAMPLE 3

How many lines of symmetry does the square have?

STRATEGY **Find the number of ways a square can be folded on a line and have both sides match exactly.**

STEP 1 Draw a horizontal line through the middle of the square.

The line makes both sides match. It is a line of symmetry.

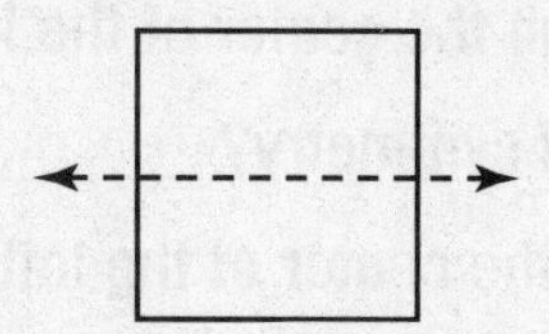

STEP 2 Draw a vertical line through the middle of the square.

The line makes both sides match. It is a line of symmetry.

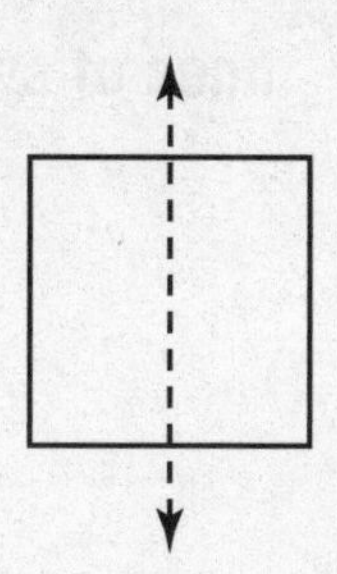

STEP 3 Check for more lines of symmetry.

Draw two diagonal lines.

The diagonals makes both sides match.

The diagonals are lines of symmetry.

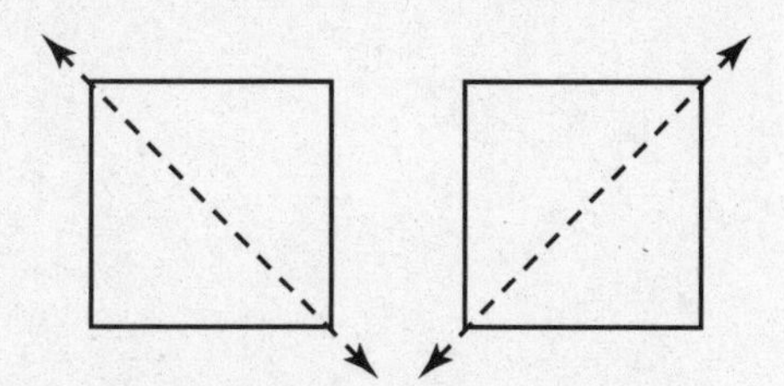

SOLUTION **The square has 4 lines of symmetry.**

COACHED EXAMPLE

How many lines of symmetry does the block letter have?

THINKING IT THROUGH

Draw a horizontal line through the center of the letter.

Is the horizontal line a line of symmetry? __________

Draw a vertical line through the center of the letter.

Is the vertical line a line of symmetry? __________

Do you see any other lines of symmetry? __________

The block letter has __________ lines of symmetry.

Lesson Practice

Choose the correct answer.

1. Which figure shows a line of symmetry?

A.

B.

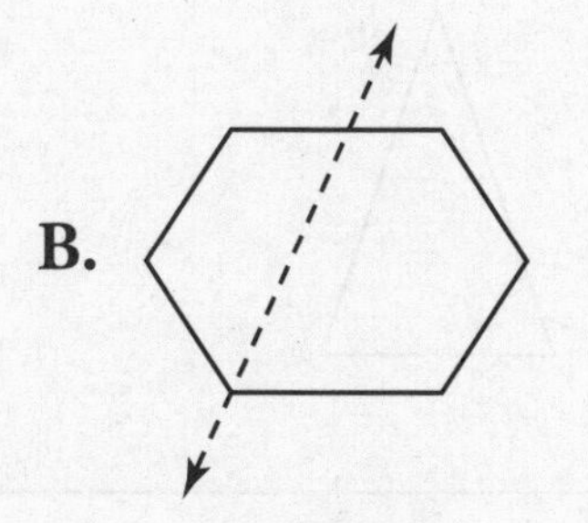

C.

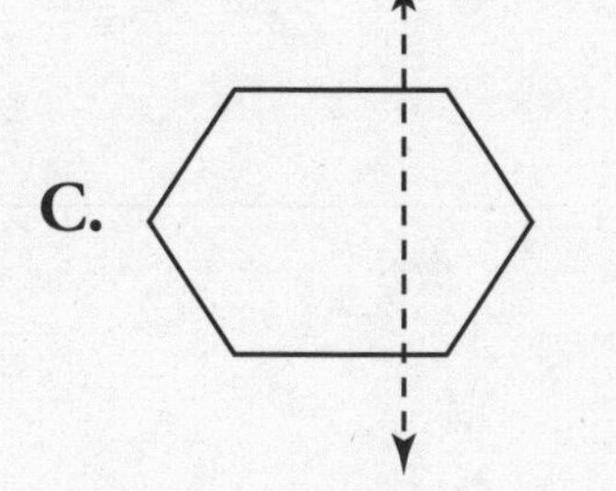

D.

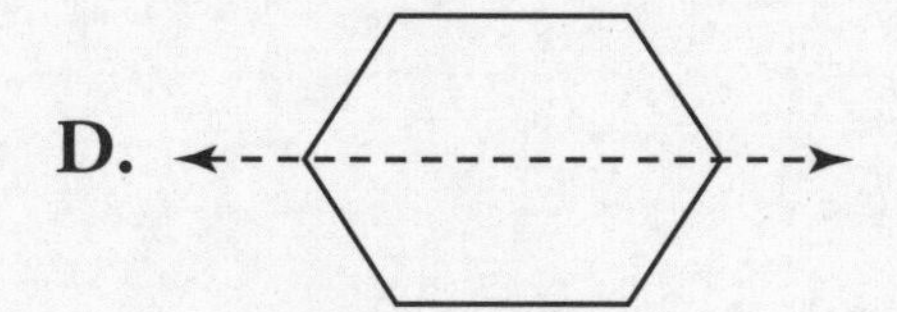

2. Which figure has **no** lines of symmetry?

A.

B.

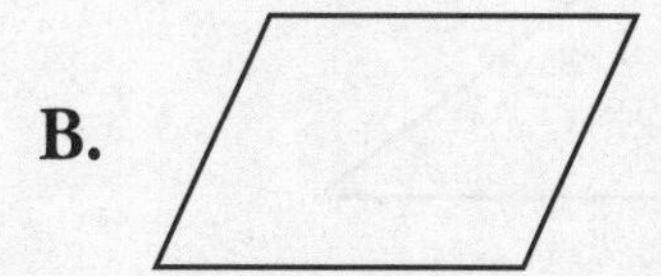

C.

D.

3. Which does **not** show a correct line of symmetry?

A.

B.

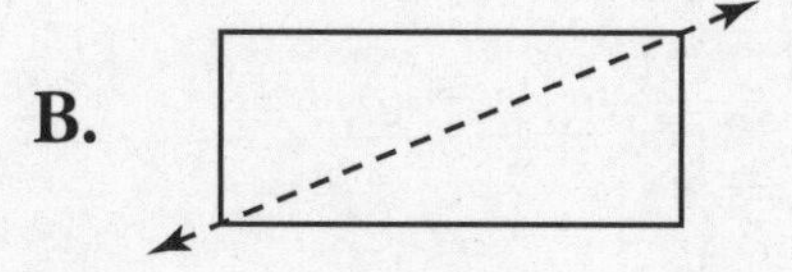

C.

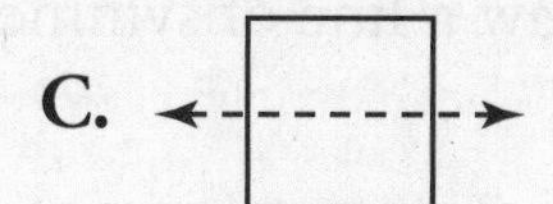

D.

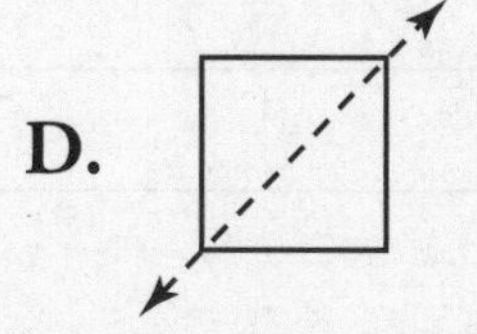

4. Which triangle has exactly 3 lines of symmetry?

A. 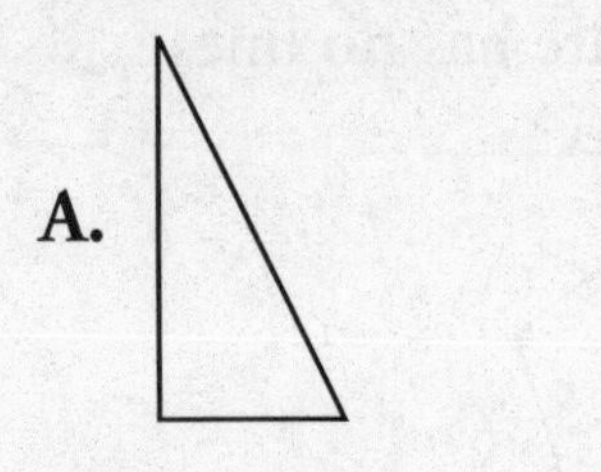

B. 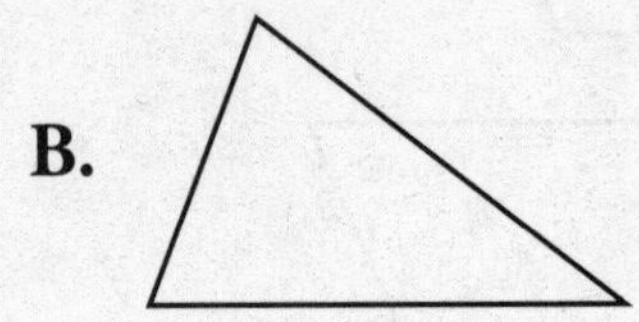

C. 

D. 

5. How many lines of symmetry does this arrow have?

Answer ______________________

6. How many lines of symmetry does this triangle have?

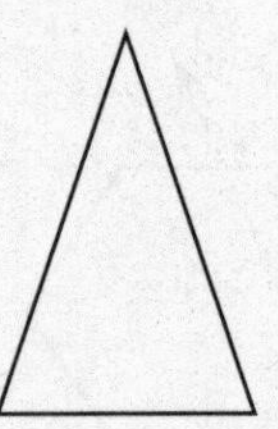

Answer ______________________

EXTENDED-RESPONSE QUESTION

7. Simon drew this figure.

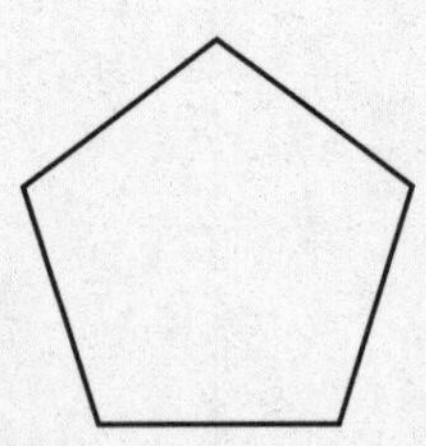

Part A Draw a line of symmetry on the figure.

Part B Does Simon's figure have symmetry? Explain how you know.

____

____

3 Review

1 Courtney drew a two-dimensional shape that has 4 sides that are the same length. Which shape did she draw?

A rhombus

B triangle

C circle

D trapezoid

2 Which two-dimensional shape has the greatest number of sides?

A square

B hexagon

C triangle

D circle

3 Which two-dimensional shape has no angles?

A circle

B triangle

C rectangle

D square

4 Which pair of shapes is congruent?

A

B

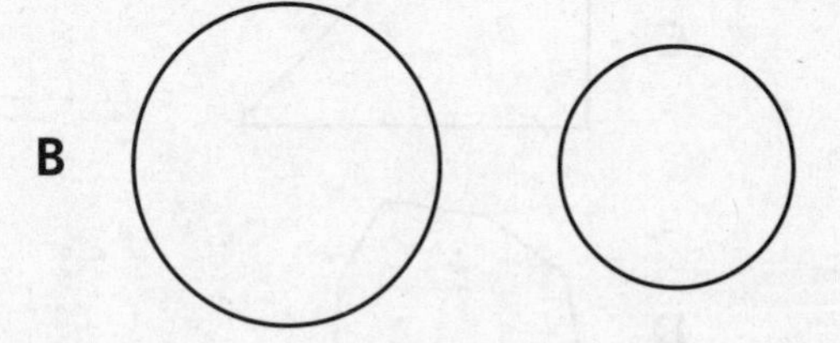

C

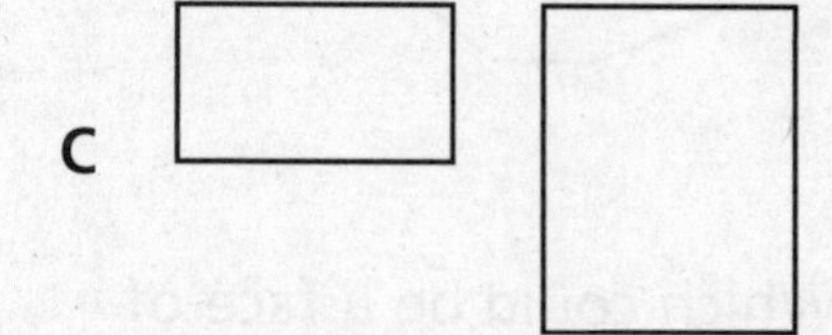

D

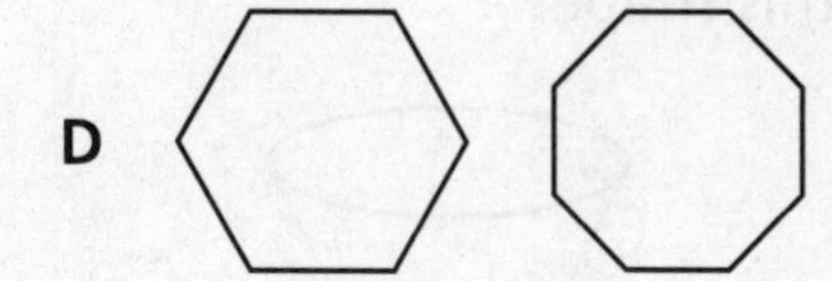

5 What is the name of this shape?

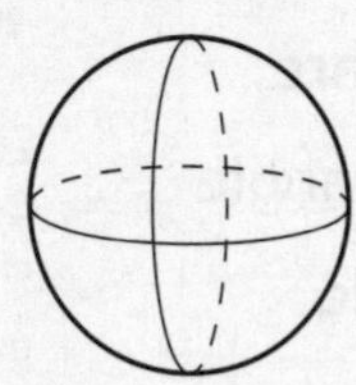

A cylinder

B cone

C sphere

D cube

6 Which is a hexagon?

A 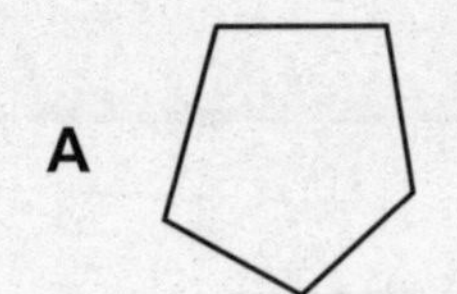

B 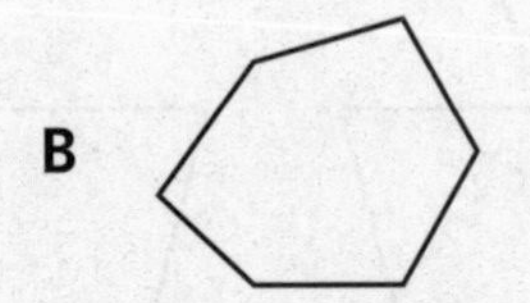

C 

D 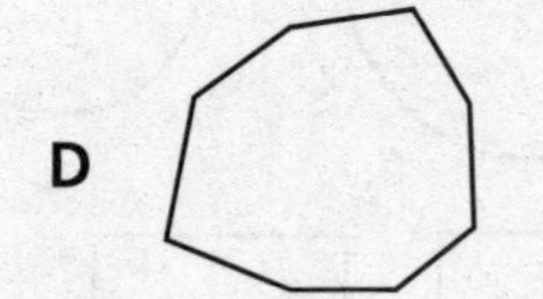

7 Which could be a face of this shape?

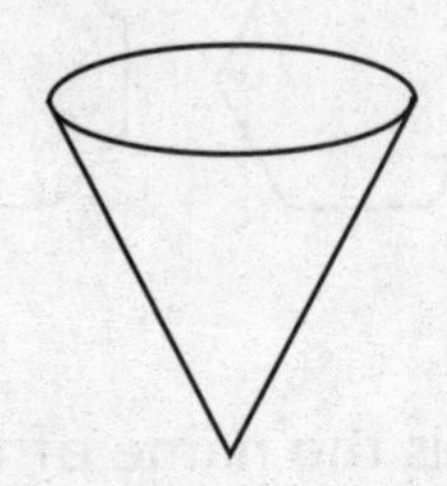

A square

B rectangle

C circle

D trapezoid

8 Which is a trapezoid?

A 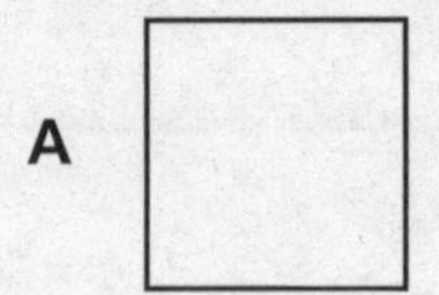

B 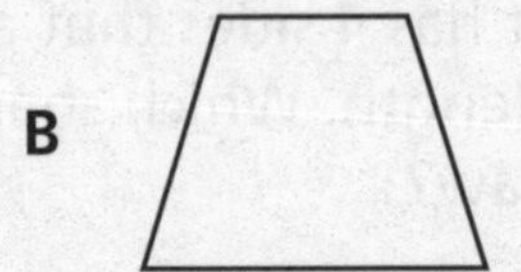

C 

D 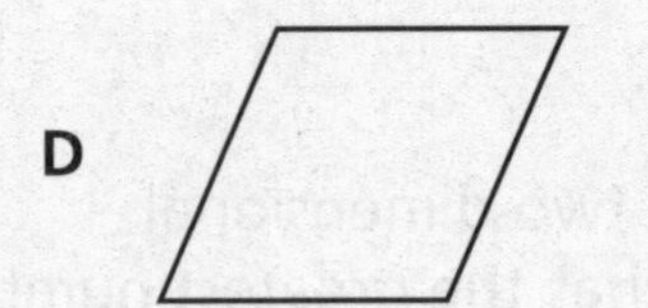

9 Which shape does not have any flat faces?

A 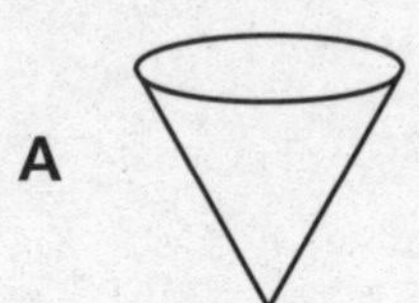

B 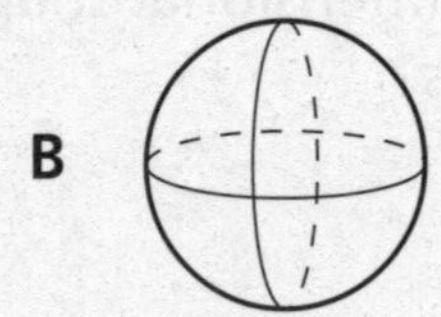

C 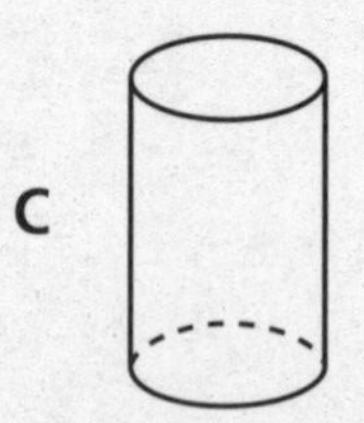

D 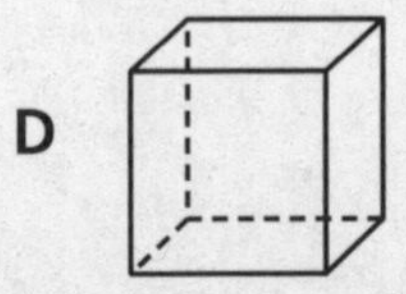

10 Which three-dimensional shape can have 2 square faces and 4 rectangular faces?

A rectangular prism

B triangular prism

C cube

D sphere

11 Which shows the correct line of symmetry?

A 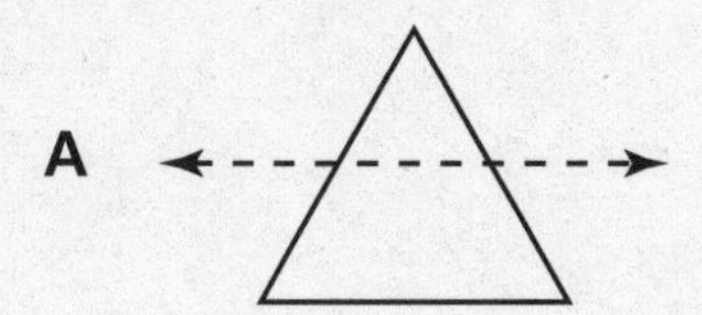

B 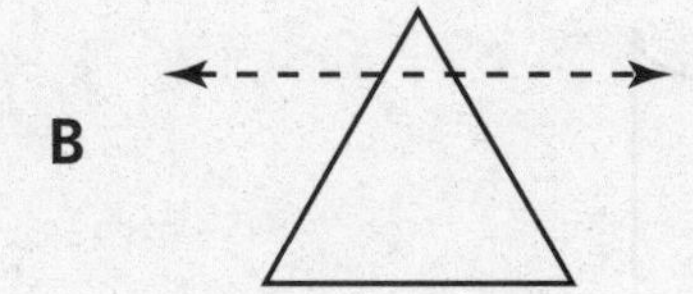

C 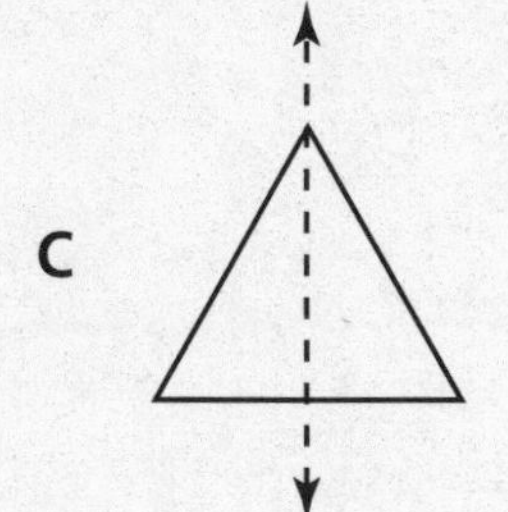

D 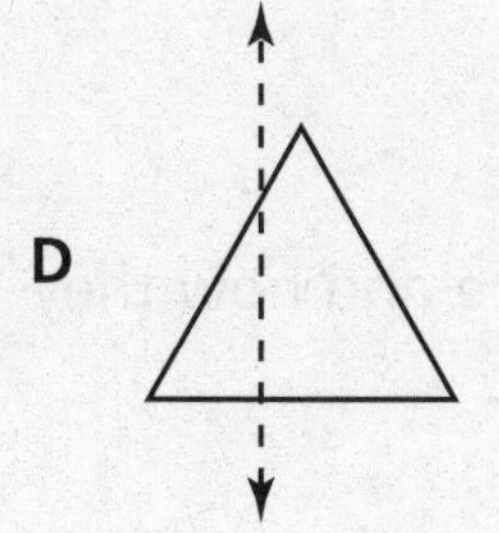

12 How many lines of symmetry does a square have?

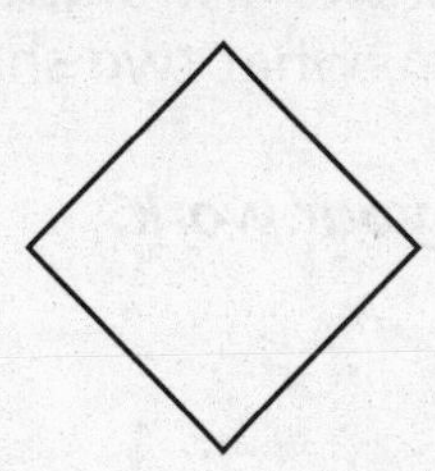

A 0

B 1

C 2

D 4

13 Pablo drew a rectangle that is 2 inches long and 1 inch wide.

If he draws a straight line in the middle of the rectangle, from the top to the bottom, what two shapes will he form?

Show your work.

Answer ____________________

14 Peggy has these two three-dimensional shapes.

Part A

What are the names of Peggy's shapes?

Answer ____

Part B

On the lines below, explain how the two shapes are alike and how they are different.

____

____

STRAND

4 Measurement

Lesson

25 Length

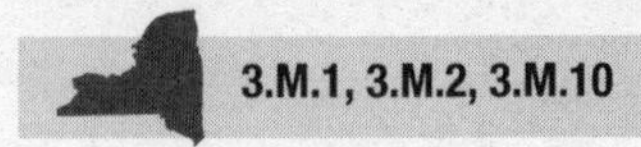

Getting the Idea

Length is the measure of how long, wide, or tall something is.

An **inch (in.)** is a unit of length in the customary system.

You can use a ruler to measure objects to the nearest inch or $\frac{1}{2}$ inch.

EXAMPLE 1

To the nearest inch, what is the length of this pencil?

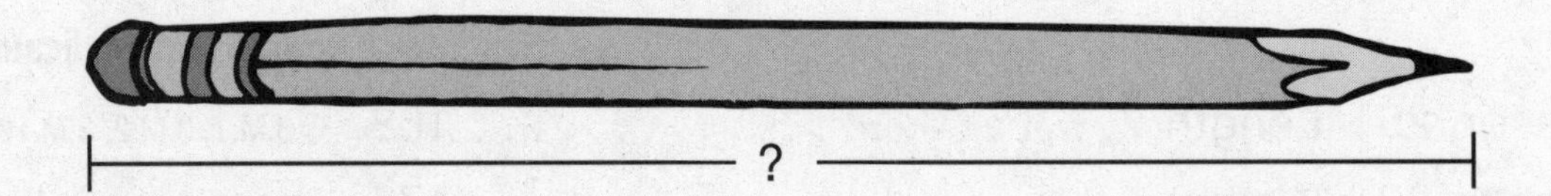

STRATEGY **Use an inch ruler.**

STEP 1 Line up the left end of the pencil with the 0 mark on the ruler.

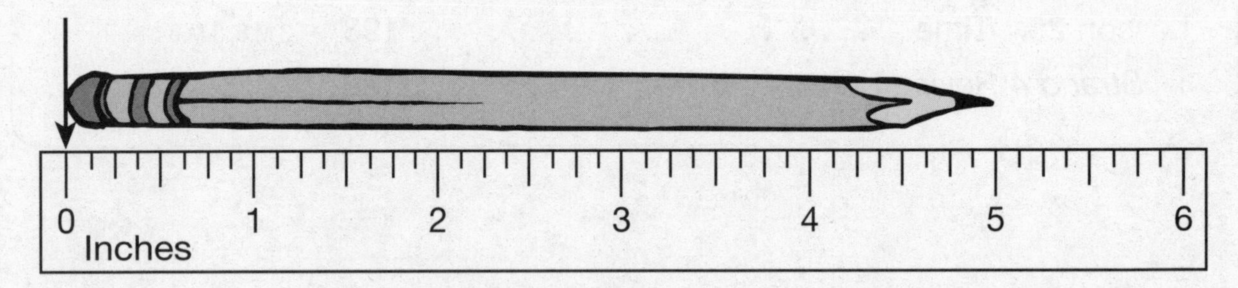

STEP 2 Read the mark on the ruler that lines up with the right end of the pencil. Read the number.

The right end of the pencil lines up with the 5-inch mark.

SOLUTION **To the nearest inch, the length of the pencil is 5 inches.**

EXAMPLE 2

To the nearest $\frac{1}{2}$ inch, what is the length of this crayon?

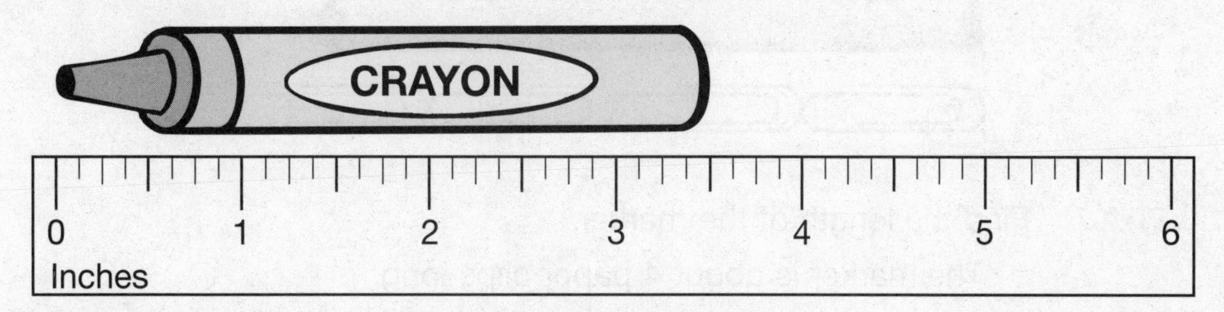

STRATEGY **Read the marks on the inch ruler.**

STEP 1 The left end of the crayon is lined up with the 0 mark on the ruler.

STEP 2 Read the mark on the ruler that lines up with the right end of the crayon.

The right end of the crayon lines up with a mark between the 3 and 4. It is halfway between 3 and 4.

The pen is $3\frac{1}{2}$ inches long.

SOLUTION **To the nearest $\frac{1}{2}$ inch, the crayon is $3\frac{1}{2}$ inches long.**

EXAMPLE 3

The paper clip is about 1 inch long. About how many inches long is the marker?

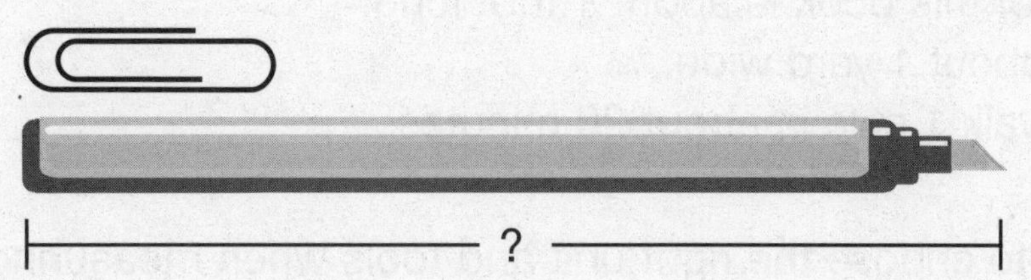

STRATEGY **Use the paper clip to estimate the length of the marker.**

STEP 1 Estimate how many paper clips long the marker is.

Imagine laying the paper clips end to end along the pencil.

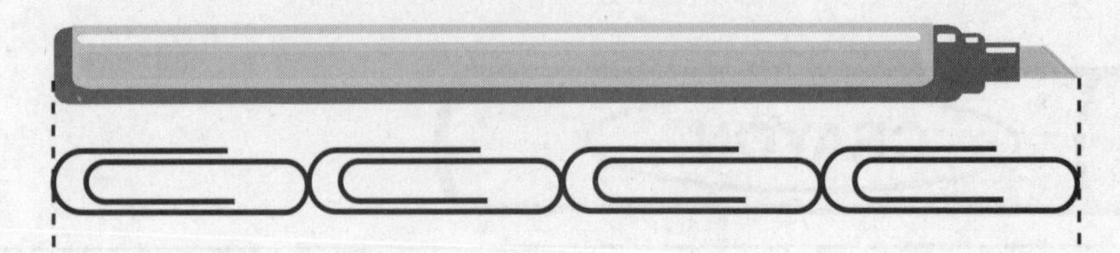

STEP 2 Find the length of the marker.

The marker is about 4 paper clips long.

Each paper clip measures about 1 inch.

So, 4 paper clips measure about 4 inches.

SOLUTION **The marker is about 4 inches long.**

The customary system has other units of length for measuring larger objects or measuring greater distances. The table shows the relationships of the units.

Customary Units of Length
12 inches = 1 **foot (ft)**
3 feet = 1 **yard (yd)**
1,760 yards = 1 **mile (mi)**

You can use these benchmarks to estimate lengths.

This line is 1 inch long. ______
The long side of this book is about 1 foot long.
A doorway is about 1 yard wide.
An adult can walk 1 mile in about 20 minutes.

It is important to choose the right unit and tools when measuring an object.

An inch ruler measures small objects that are up to 12 inches.
A yardstick measures large objects that are up to 3 feet.
A tape measure measures large objects with greater lengths.
An odometer in a car measures miles.

EXAMPLE 4

Which would be the best unit to use to measure the length of a football field?

inches feet yards miles

STRATEGY **Use benchmarks to choose the best unit.**

STEP 1 Look at the 1-inch line. It would take too many 1-inch lines to measure the football field.

STEP 2 This math book is about 1 foot long. It would take too many books to measure the football field.

STEP 3 A yard is about the width of a doorway. Using this length makes sense.

STEP 4 A mile is used to measure long distances. A mile is too long to measure a football field.

SOLUTION **Yards would be the best unit to use to measure a football field.**

EXAMPLE 5

Which is the best tool to measure the length of a chalkboard eraser?

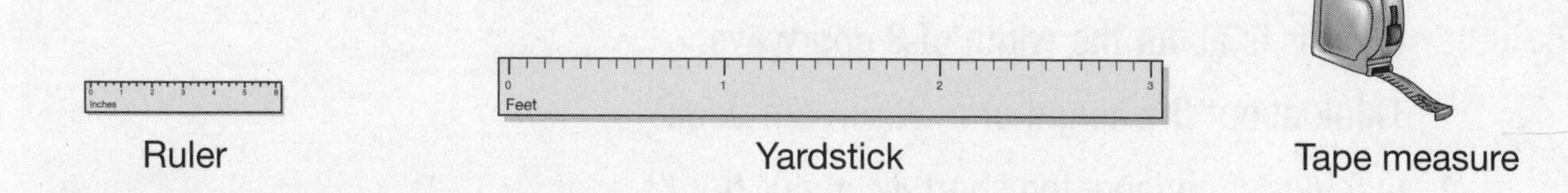

Ruler Yardstick Tape measure

STRATEGY **Think about the size of a chalkboard eraser.**

STEP 1 Is a chalkboard eraser more than 12 inches?

A chalkboard eraser is a small object.

STEP 2 Which tool do you use to measure small objects?

Use a ruler to measure small objects up to 12 inches.

SOLUTION **A ruler is the best tool to use to measure the length of a chalkboard eraser.**

COACHED EXAMPLE

Which is the best estimate for the height of a classroom door?

8 inches 8 feet 8 yards 8 miles

THINKING IT THROUGH

Compare 8 inches to a benchmark.

Use an inch ruler to draw a line that is 1 inch long.

8 inches is the length of eight 1-inch lines.

Think about the height of a classroom door.

Is 8 inches too long, too short, or about right? ______________

Compare 8 feet to a benchmark.

The long side of this math book is about ____________.

8 feet is about the length of 8 books.

Think about the height of a classroom door.

Is 8 feet too long, too short, or about right? ______________

Compare 8 yards to a benchmark.

The width of a doorway is about ____________.

8 yards is about the width of 8 doorways.

Think about the height of a classroom door.

Is 8 yards too long, too short, or about right? ______________

Compare 8 miles to a benchmark.

An adult can walk ____________ in about 20 minutes.

Think about the height of a classroom door.

Is 8 miles too long, too short, or about right? ______________

The best estimate for the height of a classroom door is ________________.

Lesson Practice

Choose the correct answer.

1. Use an inch ruler. To the nearest inch, what is the length of this pen?

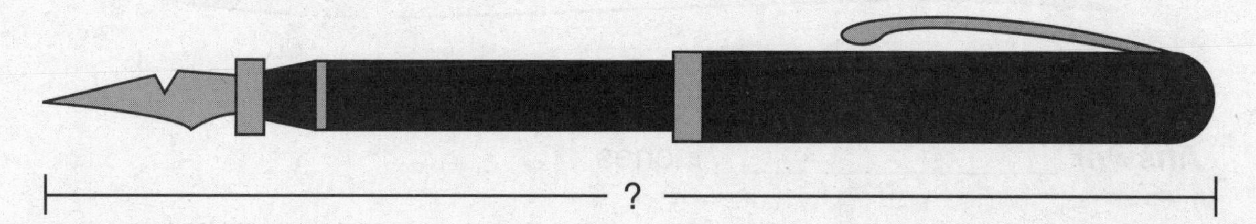

A. 4 in.

B. 5 in.

C. 6 in.

D. 7 in.

2. Which is the best tool to use to measure the length of your bedroom?

A. ruler

B. paper clip

C. tape measure

D. odometer

3. Which is the best estimate for the length of a toothbrush?

A. 6 in.

B. 60 in.

C. 6 ft

D. 60 ft

4. Hayley walked for 25 minutes in Central Park. Which is the distance she most likely walked?

A. 1 inch

B. 1 foot

C. 1 yard

D. 1 mile

5. Which is the best estimate for the length of a piece of celery?

A. 7 inches

B. 7 feet

C. 7 yards

D. 7 miles

6. The gray tile is 1 inch long. About how long is the ribbon?

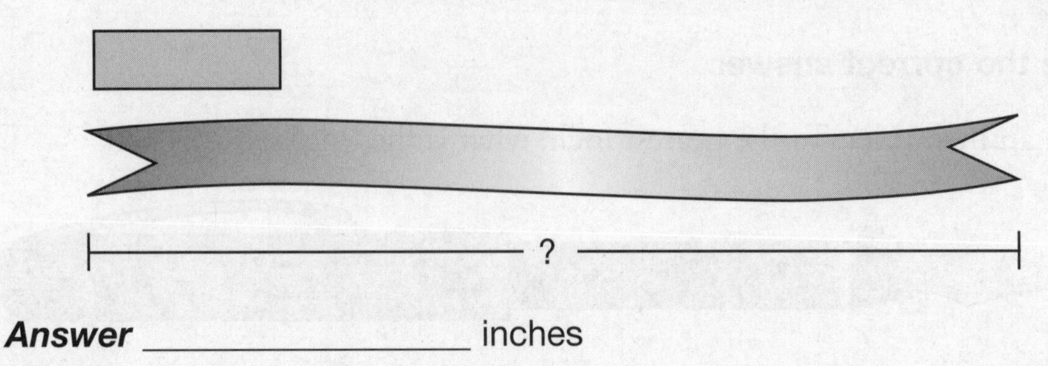

Answer ________________ inches

EXTENDED-RESPONSE QUESTION

7. Victor drew an arrow on a poster.

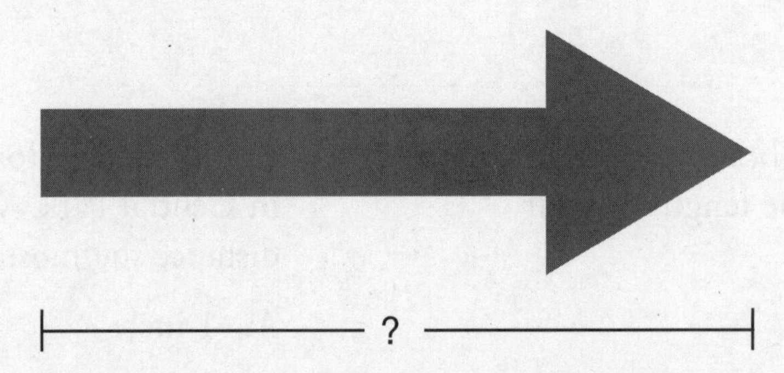

Part A Use an inch ruler. To the nearest $\frac{1}{2}$ inch, how many inches long is the arrow?

____

____

Part B Use the space below to draw an arrow that is $\frac{1}{2}$ inch longer than Victor's arrow.

Lesson

26 Weight

Getting the Idea

Weight is a measure of how heavy something is.

The table shows units of weight in the customary system.

Customary Units of Weight
16 **ounces (oz)** = 1 **pound (lb)**

You can use these benchmarks to estimate weights.

A pencil weighs about 1 ounce.

A sneaker weights about 1 pound.

The tool used to measure weight is a scale.

Below are different types of scales.

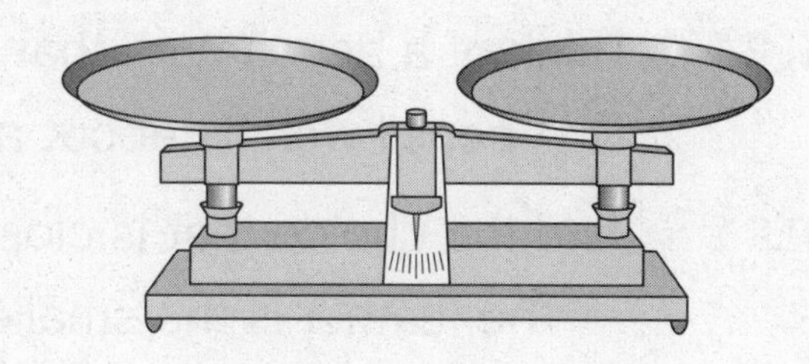

EXAMPLE 1

About how much does the broccoli weigh?

STRATEGY **Use the scale.**

STEP 1 Look at the scale.

The scale is in pounds.

STEP 2 Read the number the needle is pointing to.

The needle points close to the 2.
It shows about 2 pounds.

SOLUTION **The broccoli weighs about 2 pounds.**

EXAMPLE 2

Which has a weight best measured in ounces?

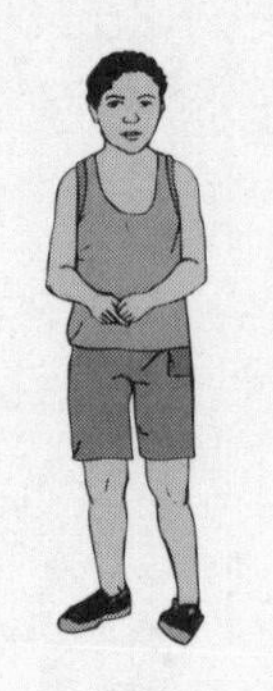

Feather Boy Chair Elephant

STRATEGY **Compare each choice to 1 ounce.**

STEP 1 Think of a benchmark that weighs 1 ounce.

A pencil weighs about an ounce.

STEP 2 Find the choice that is closest to the weight of a pencil.

The feather is the smallest and lightest.

STEP 3 Check the other choices.

The boy, chair, and elephant all weigh much more than 1 pound.

Their weights would best be measured in pounds.

SOLUTION **The weight of the feather is best measured in ounces.**

EXAMPLE 3

Which is the best estimate for the weight of a bicycle?

2 ounces   2 pounds   20 ounces   20 pounds

STRATEGY **Use benchmarks to find the best estimate.**

STEP 1 Think about some benchmarks.

A pencil weighs about 1 ounce.

A sneaker weighs about 1 pound.

STEP 2 Review the choices.

2 ounces is about the weight of 2 pencils. Too light.

2 pounds is about the weight of 2 sneakers. Too light.

20 ounces is about the weight of 20 pencils. Too light.

20 pounds is about the weight of 20 sneakers. This makes sense.

SOLUTION **The best estimate for the weight of a bicycle is 20 pounds.**

COACHED EXAMPLE

Which fruit most likely weighs about 7 ounces?

cherry

strawberry

watermelon

banana

THINKING IT THROUGH

Think of a benchmark that weighs about 1 ounce. ____________________

So, 7 _______________ weigh about 7 ounces.

Check the choices.

The cherry weighs less than 7 ounces.

The _______________ weighs less than 7 ounces.

The _______________ weighs more than 7 ounces.

Which fruit weighs about 7 ounces? _______________

The _______________ most likely weighs 7 ounces.

Lesson Practice

Choose the correct answer.

1. Which is the best unit to use to estimate the weight of an 18-wheel truck?

A. ounces

B. inches

C. pounds

D. miles

2. About how much do the books weigh?

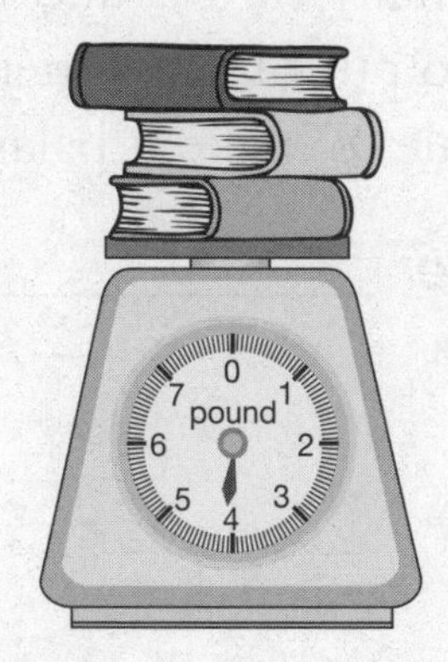

A. 3 pounds

B. 3 ounces

C. 4 pounds

D. 4 ounces

3. Which most likely weighs about 50 pounds?

A. 

B. 

C. 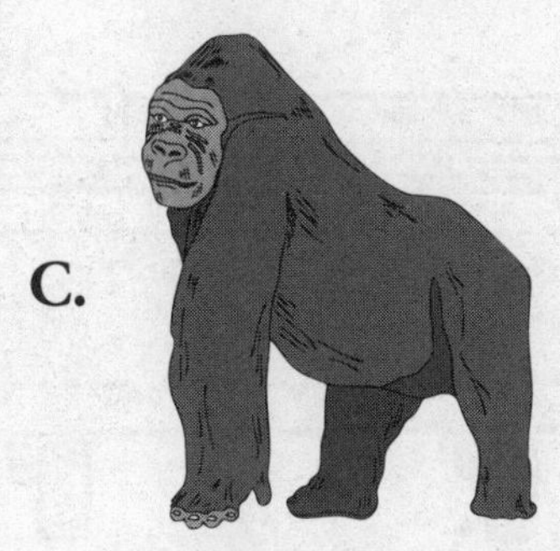

D. 

4. Which is best estimate for the weight of a cotton ball?

A. 1 ounce

B. 10 ounces

C. 1 pound

D. 10 pounds

5. Which is the best estimate for the weight of a cell phone?

A. 800 ounces

B. 80 pounds

C. 8 pounds

D. 8 ounces

6. Which object is best measured in pounds?

A. 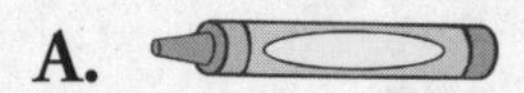

B. 

C.

D.

7. Fill in the blank with a unit of weight.

A watermelon weighs about 4 ________.

Answer ________________

8. George wants to balance the scale. The cauliflower weighs about 1 pound. Each carrot weighs about 4 ounces.

4 ounces

How many carrots does George need to put on the other side of the scale to make it balance?

Answer ________________

Lesson

27 Capacity

3.M.4, 3.M.5, 3.M.6, 3.M.10

Getting the Idea

Capacity is the measure of how much a container can hold.

The table shows the units of capacity in the customary system.

Customary Units of Capacity
2 **cups (c)** = 1 **pint (pt)**
2 pints = 1 **quart (qt)**
4 quarts = 1 **gallon (gal)**

Look at the containers below.

You can use these benchmarks to estimate capacity.

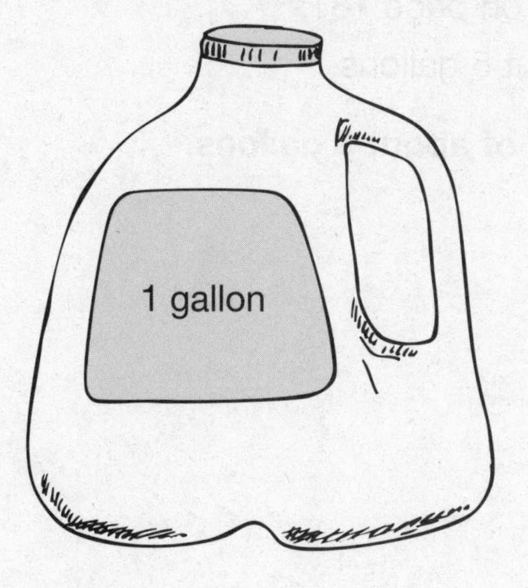

EXAMPLE 1

Does a kitchen sink have a capacity of about 5 cups or 5 gallons?

STRATEGY **Use benchmarks to find the best estimate.**

STEP 1 Think about 5 cups.

Look at the cup container on page 181.

A kitchen sink can hold much more than 5 cups.

STEP 2 Think about 5 gallons.

Look at the gallon container on page 181.

A kitchen sink can hold about 5 gallons.

SOLUTION **A kitchen sink has a capacity of about 5 gallons.**

EXAMPLE 2

Joel needs 1 quart of chicken stock for a recipe. He only has a 1-cup measuring cup. How many times will Joel need to fill the measuring cup to have 1 quart of chicken stock?

STRATEGY **Think about the relationship between the two units.**

STEP 1 What is the relationship between a quart and a cup?

A quart is larger than a cup.

4 cups = 1 quart

STEP 2 Find how many times Joel has to fill the cup.

The measuring cup has a capacity of 1 cup.

4 cups make a quart.

SOLUTION **Joel has to fill the measuring cup 4 times to have 1 quart of chicken stock.**

EXAMPLE 3

Order the containers in order from least capacity to greatest capacity.

juice box   bathtub   cereal bowl

STRATEGY **Use benchmarks to estimate each capacity.**

STEP 1 Compare a juice box to the containers on page 181.

A juice box holds about 1 cup.

STEP 2 Compare a bathtub to the containers on page 181.

A bathtub holds more than 1 gallon.

STEP 3 Compare a cereal bowl to the containers on page 181.

A cereal bowl holds about 1 pint.

STEP 4 Order the capacities.

1 cup, 1 pint, more than 1 gallon

SOLUTION **The order of containers from least capacity to greatest capacity is juice box, cereal bowl, and bathtub.**

COACHED EXAMPLE

Terry wants to put 2 quarts of water in a bucket. He only has a measuring cup that holds 1 pint. How can Terry fill the bucket with 2 quarts of water?

THINKING IT THROUGH

Is a quart larger or smaller than a pint? _______________

Terry needs to keep filling the 1-pint measuring cup.

1 quart = _____ pints

2 quarts = _____ pints

How many times does Terry have to fill the measuring cup? _______

Terry can fill the 1-pint measuring cup _____ times to measure 2 quarts.

Lesson Practice

Choose the correct answer.

1. Which is the best estimate for the capacity of a bathtub?

A. 5 cups

B. 50 pints

C. 5 quarts

D. 50 gallons

2. Which would be best measured in gallons?

A. the amount of milk that you drink at breakfast

B. the amount of juice a drinking glass holds

C. the amount of liquid medicine you take

D. the amount of water in a fish tank

3. Which is **not** a unit of capacity?

A. pint

B. pound

C. quart

D. gallon

4. Which is the best estimate for the capacity of a glass of iced tea?

A. 1 pint

B. 10 pints

C. 1 quart

D. 10 quarts

5. Which sentence is true?

A. A cup is greater than a pint.

B. A quart is less than a gallon.

C. A pint is greater than a quart.

D. A gallon is less than a quart.

6. Which object has the greatest capacity?

   A. a soda can
   B. a medicine dropper
   C. a washing machine
   D. an inground swimming pool

7. Which is most likely to have a capacity of 1 pint?

   A. a scoop of ice cream
   B. a bowl of soup
   C. a gas tank
   D. a handful of water

8. Harold poured some milk into a bowl with cereal. Which measure is **most likely** the amount of milk that Harold poured in the cereal?

   A. 1 cup
   B. 10 cups
   C. 1 quart
   D. 10 quarts

9. Ling wants to measure 2 pints of milk for a recipe. She only has a 1-cup measuring cup. How many times must Ling fill the 1-cup measuring cup to measure 2 pints?

   Answer ______________

10. Jennifer poured 3 gallons of hot water into the tub. How many quarts of hot water did Jennifer pour into the tub?

   Answer ______________ quarts

Lesson

28 Money

Getting the Idea

Money is used to buy things. Below are the coins and bills that are used most often in the United States.

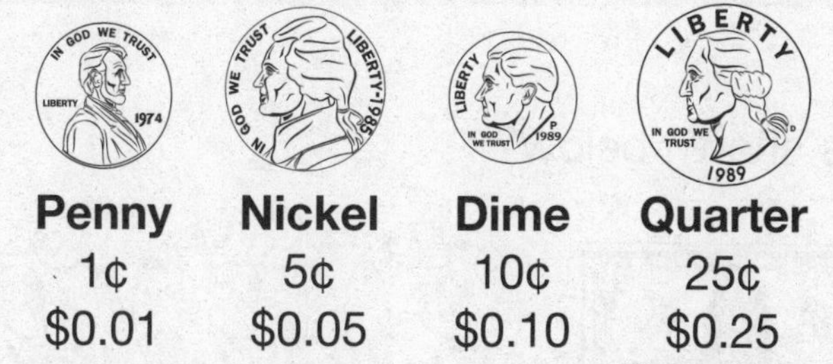

Penny	Nickel	Dime	Quarter
1¢	5¢	10¢	25¢
$0.01	$0.05	$0.10	$0.25

One-Dollar Bill
$1.00 or $1

Five-Dollar Bill
$5.00 or $5

To find the value of a group of coins, count on from the coin with the greatest value to the coin with the least value.

EXAMPLE 1

Robert has these coins in his pocket. How much money does Robert have?

STRATEGY **Sort the coins by value. Count on from the greatest value to the least value.**

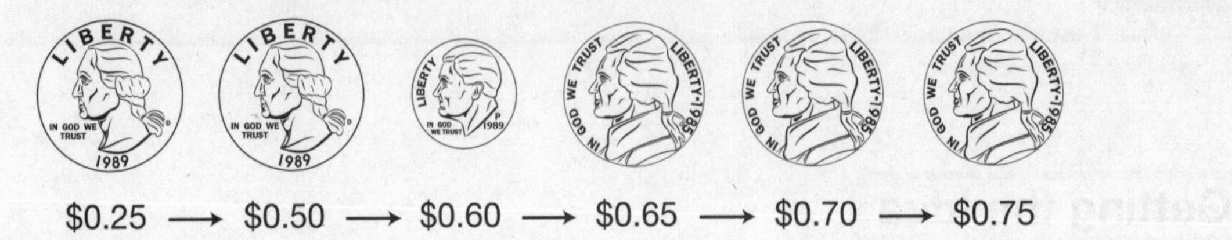

$0.25 ⟶ $0.50 ⟶ $0.60 ⟶ $0.65 ⟶ $0.70 ⟶ $0.75

SOLUTION **Robert has $0.75.**

EXAMPLE 2

How much money is shown below?

STRATEGY **Count the bills first. Then count on the coins from the greatest value to the least value.**

STEP 1 Count on the bills.

$1 → $2

STEP 2 Count on the coins from the greatest value to least value.

Start at $2.

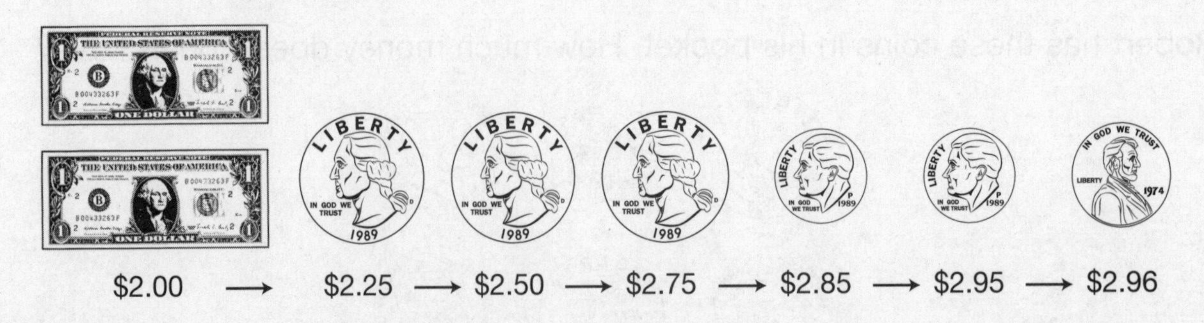

$2.00 ⟶ $2.25 ⟶ $2.50 ⟶ $2.75 ⟶ $2.85 ⟶ $2.95 ⟶ $2.96

SOLUTION **Christine has $2.96.**

EXAMPLE 3

The amount of money Leah and Brandon have is shown below.

Who has more money?

STRATEGY **Count the bills and coins in each group. Compare the amounts.**

STEP 1 Count Leah's money.

bills: $\$1 \rightarrow \$2 \rightarrow \$3$

coins: $\$3.25 \rightarrow \$3.50 \rightarrow \$3.55 \rightarrow \$3.60 \rightarrow \$3.61$

STEP 2 Count Brandon's money.

bills: $\$1 \rightarrow \$2 \rightarrow \$3$

coins: $\$3.25 \rightarrow \$3.35 \rightarrow \$3.45 \rightarrow \$3.55 \rightarrow \$3.60 \rightarrow \$3.65$

STEP 3 Compare the amounts.

$3.61 is less than $3.65.

SOLUTION **Brandon has more money.**

COACHED EXAMPLE

List two different ways that you could have $0.85 in coins.

THINKING IT THROUGH

Think about the value of each type of coin.

A quarter is worth _______ cents.

A dime is worth _______ cents.

A nickel is worth _______ cents.

A penny is worth _______ cent.

List one way that you could have $.0.85 with coins.

____

____

____

List a different way that you could have $.0.85 with coins.

____

____

____

Lesson Practice

Choose the correct answer.

1. Isabella found $0.46 under the sofa cushions. Which shows the money Isabella found?

A. 

B. 

C. 

D. 

2. What is the value of the money shown?

A. $1.60

B. $1.65

C. $1.70

D. $1.75

3. Paul has these coins in a drawer.

Which group of coins has a value greater than Paul's coins?

A. 

B. 

C. 

D. 

4. Which group of coins shows $0.55?

A. 

B. 

C. 

D.

5. Susan has these bills and coins to buy a pen. How much does Susan have?

Answer ______________

6. The bills and coins below show the cost of a notebook. How much does the notebook cost?

Answer ______________

EXTENDED-RESPONSE QUESTION

7. Keisha has these bills and coins in a jar.

How much money is in the jar?

____

8. List two different ways you can make $0.99 with coins.

____

____

____

Lesson

29 Time

Getting the Idea

A clock is used to measure time. On an analog clock, the short hand points to the **hour**. The long hand points to the **minute**.

The minute hand moves to the next number every 5 minutes.
The hour hand moves to the next number every 60 minutes, or 1 hour.

12:00
twelve
o'clock

12:15
a quarter
past twelve

12:30
half past
twelve

12:45
a quarter
to one

Units of Time
15 minutes = $\frac{1}{4}$ hour
30 minutes = $\frac{1}{2}$ hour
45 minutes = $\frac{3}{4}$ hour
60 minutes = 1 hour

EXAMPLE 1

The clock shows the time that Aiden's bus picks him up for school.
What time does Aiden's bus pick him up?

STRATEGY **Look at the hands of the clock.**

STEP 1 Look at the shorter hand to tell the hour.

The hour hand is between 8 and 9, so it is after 8 o'clock.

STEP 2 Look at the longer hand to tell the minutes.

There are 5 minutes between each number.

The minute hand points to the 3.

It is 15 minutes, or $\frac{1}{4}$ of an hour, past the hour.

SOLUTION **Aiden's bus picks him up at 8:15.**
Read: eight fifteen, or a quarter past eight o'clock

EXAMPLE 2

The clock shows the time that Margie got home from school today.
What time did Margie get home from school?

STRATEGY **Look at the hands of the clock.**

STEP 1 Look at the hour hand.

It is between 3 and 4, so it is after 3 o'clock.

STEP 2 Look at the minute hand.

When the minute hand points to the 7, it is 35 minutes past the hour. The minute hand points to the second mark after 7. Count forward 2 minutes from 35.

It is 37 minutes past 3 o'clock.

SOLUTION **Margie got home at 3:37 today.**
Read: three thirty-seven or thirty-seven minutes past three o'clock

COACHED EXAMPLE

The clock shows the time that a plane landed at the airport. What time did the plane land?

THINKING IT THROUGH

Look at the hour hand.

The hour hand is between _______ and ______.

The time is past _____ o'clock.

Look at the minute hand.

Count the minutes by fives. Then count on by ones.

It is ______ minutes past the hour.

The plane landed at _____:_____.

Read the time as ____.

Lesson Practice

Choose the correct answer.

1. The clock shows the time that Quincy finished football practice.

What time did Quincy finish football practice?

A. 5:28

B. 5:43

C. 6:43

D. 8:28

2. Which digital clock shows the same time as the analog clock?

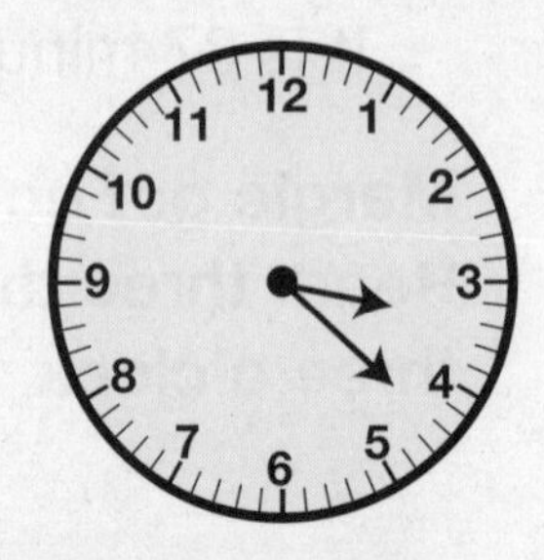

A. 3:22

B. 3:40

C. 3:50

D. 4:15

3. Which is equal to 30 minutes?

A. 1 hour

B. $\frac{1}{4}$ hour

C. $\frac{1}{2}$ hour

D. $\frac{3}{4}$ hour

4. Rosa's ballet class starts at 4:15. Which is **not** a way to read the time when the ballet class starts?

   A. half past four

   B. a quarter past four

   C. fifteen minutes after four

   D. four fifteen

5. The clock shows the time that Marta started cooking lunch. Which shows the way to read the time?

   A. fifteen minutes past twelve

   B. a quarter to twelve

   C. a quarter past eleven

   D. twelve minutes after nine

6. Look at this digital clock.

What is the time in words?

Answer ____________

7. What time does the clock show?

Answer ____________

EXTENDED-RESPONSE QUESTION

8. The clock shows the time that the students have gym class.

Part A What time do the students have gym class?

____

Part B Use words to write the time in 2 different ways.

____

____

Part C Explain how you solved each part of the problem.

____

____

____

4 Review

1  Use your ruler to help you solve this problem.

How many inches long is the key shown below?

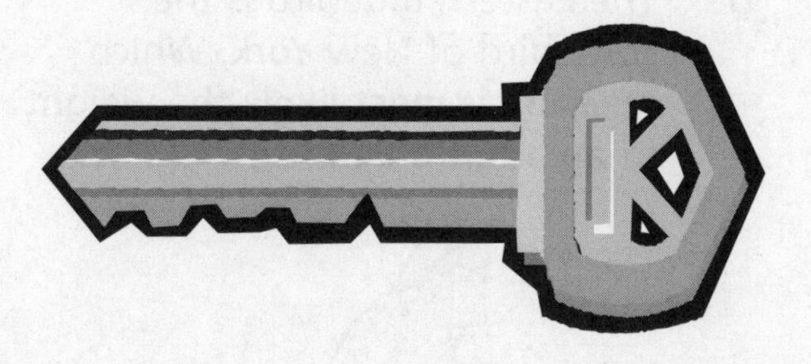

A $2\frac{1}{2}$ in.

B 3 in.

C $3\frac{1}{2}$ in.

D 4 in.

2  Use your ruler to help you solve this problem.

The drawing below shows a drawing of Kelly's stuffed animal.

What is the height of the drawing?

A $1\frac{1}{2}$ in.

B 2 in.

C $2\frac{1}{2}$ in.

D 3 in.

3 Kyle is using an inch ruler. Which object is Kyle **most likely** measuring?

A the height of a sliding door

B the width of a sectional couch

C the length of a big-screen TV

D the height of a drinking glass

4 Which object weighs about 1 ounce?

A 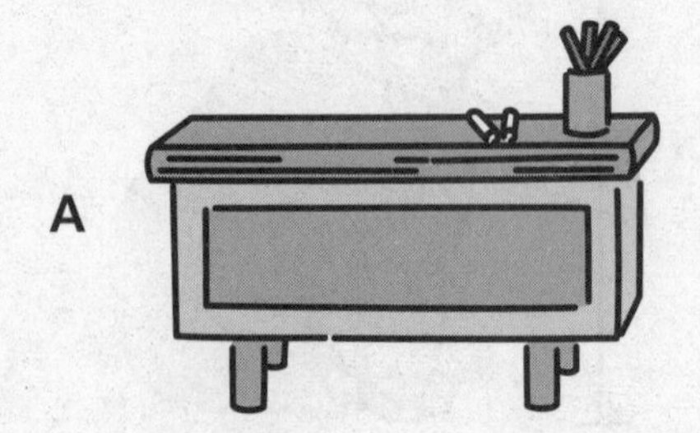

B 

C 

D 

5 Which is **most likely** the weight of a basketball?

A 2 ounces

B 20 ounces

C 10 pounds

D 20 pounds

6 The Eastern Bluebird is the state bird of New York. Which measure is **most likely** the weight of an adult bluebird?

A 1 ounce

B 10 pounds

C 100 ounces

D 100 pounds

7 Which item **most likely** would weigh about 40 pounds?

A a cat

B a car

C a desk

D an apple

8 Mr. Murray made a pitcher of lemonade. Which is the **best** estimate for the amount of lemonade he made?

A 2 cups

B 2 quarts

C 4 gallons

D 8 gallons

9 Which is the **least** measure of capacity?

A 1 quart

B 1 pint

C 1 cup

D 1 gallon

10 Mindy has $0.25 in her pocket. She needs $0.75 to buy a pen. Which shows the coins that Mindy still needs to buy the pen?

A 

B 

C 

D 

11 The clock shows the time Mark's friends arrived at his house. What time did his friends arrive?

A 9:13

B 8:13

C 8:53

D 9:07

12 Mr. Jefferson started driving at 6:30 this morning. Which is one way to read the time that he started driving?

A three sixty

B a quarter past six

C half past six

D fifteen minutes to seven

13 What time does the clock show?

Answer ________________

14 The bills and coins below show the cost of a value meal at a restaurant.

What is the cost of the value meal?

Answer ________________

15 The paper clip below is 2 inches long. Allison used this paper clip to estimate the length of the paintbrush.

?

Part A

About how many inches long is the paintbrush?

Answer ______________ inches

Part B

On the lines below, explain how you estimated the length of the paintbrush.

____

____

____

STRAND

5 Statistics and Probability

** Grade 3 May–June Indicators

30 Frequency Tables

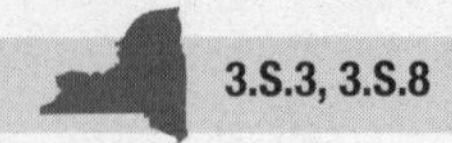

Getting the Idea

Data is information. You can sort collected data into groups.

You can use a **frequency table** to organize the data.

Some frequency tables have tally marks. In tally marks,

| represents 1 and 𝍸 represents 5.

EXAMPLE 1

Thomas recorded the numbers of meals served at a restaurant. How many more turkey meals were served than lasagna meals?

Meals Served at Restaurant

Meal	Tally
Meatloaf	𝍸 𝍸 |||
Lasagna	𝍸 𝍸
Turkey	𝍸 𝍸 𝍸	
Tacos	𝍸 𝍸 ||

STRATEGY **Count the tallies for each group.**

STEP 1 Count the number of tallies in the turkey row.

There are 16 tallies.

STEP 2 Count the number of tallies in the lasagna row.

There are 10 tallies.

STEP 3 "How many more" means to subtract.

$16 - 10 = 6$

SOLUTION **There were 6 more turkey meals served than lasagna meals.**

EXAMPLE 2

Rachel asked her classmates how many dogs they own.
She wrote the number of dogs each classmate owns below.

1 2 0 3 0 3 0 2 0 0 1

1 3 1 0 1 2 0 1 1 2 1

Make a frequency table to represent the data.

STRATEGY **Record the data in a frequency table.**

STEP 1 Make a table.

Write each of the responses given in the left column.

Use the right column to tally how many of each response.

Dogs Owned

Number of Dogs Owned	Tally
0	
1	
2	
3	

STEP 2 Record the data.

Mark a tally for each piece of data.

Cross out each piece of data as it is recorded.

Dogs Owned

Number of Dogs Owned	Tally
0	𝍸 II
1	𝍸 III
2	IIII
3	III

SOLUTION **The frequency table is shown in Step 2.**

You can sometimes make predictions based on the data in a table.

A **prediction** is an educated guess as to what will happen.

EXAMPLE 3

The table shows the grades Natalie received on her mathematics quizzes.

Natalie's Mathematics Quiz Grades

Quiz Grade	Tally
A	~~IIII~~ ~~IIII~~ II
B	II
C	I

Based on the data in the table, predict what grade Natalie will probably get on her next quiz.

STRATEGY **Analyze the data. Use what has happened to predict what will probably happen.**

STEP 1 Find how many times Natalie got each grade.

Natalie got 12 A's, 2 B's, and 1 C.

STEP 2 Compare the numbers.

She got 9 more A's than B's and C's combined.

She will probably get an A.

SOLUTION **Natalie will probably get an A on her next quiz.**

COACHED EXAMPLE

The students in Mr. Renfro's class were asked to name their favorite season. The results are listed below.

Fall	**Summer**	**Summer**	**Summer**	**Spring**
Winter	**Fall**	**Summer**	**Summer**	**Fall**
Spring	**Summer**	**Spring**	**Fall**	**Spring**
Winter	**Summer**	**Fall**	**Summer**	**Winter**
Fall	**Spring**	**Summer**	**Summer**	**Fall**

How many more students chose the most popular season than chose the least popular season?

THINKING IT THROUGH

Complete the frequency table.

How do you represent 1 using tally marks? __________

How do you represent 5 using tally marks? __________

Favorite Seasons

Season	Tally
Fall	
Winter	
Spring	
Summer	

Which is the most popular season? _______________

How many students picked the most popular season? _____

Which is the least popular season? _______________

How many students picked the least popular season? _____

Subtract. _____ – _____ = _____

______________ **more students chose the most popular season than chose the least popular season.**

Lesson Practice

Choose the correct answer.

Use the data below for questions 1–3.

The teams in a mathematics contest played 12 rounds. The list below shows the winning team in each round.

Bulldogs	Tigers	Tigers	Bobcats
Bulldogs	Bobcats	Tigers	Bobcats
Tigers	Tigers	Tigers	Bulldogs

1. Which sentence is true?

   A. The Tigers have won more times than the other two teams combined.

   B. The Bulldogs have won as many times as the Bobcats.

   C. The Bobcats have won more times than the Bulldogs.

   D. The Bulldogs have won more times than the Bobcats.

2. The Falcons also competed, but have yet to win. Based on the data, which team do you predict would win the next round?

   A. Tigers

   B. Bulldogs

   C. Bobcats

   D. Falcons

3. Which table correctly shows the data?

A.

Math Contest

Team	Tally
Bulldogs	III
Tigers	~~IIII~~ I
Bobcats	II

B.

Math Contest

Team	Tally
Bulldogs	III
Tigers	~~IIII~~
Bobcats	III

C.

Math Contest

Team	Tally
Bulldogs	III
Tigers	~~IIII~~ I
Bobcats	III

D.

Math Contest

Team	Tally
Bulldogs	III
Tigers	~~IIII~~
Bobcats	II

4. The table shows some students' favorite winter sports. How many more students picked snowboarding than sledding?

Favorite Winter Sport

Sport	Tally
Skiing	卌 卌 ||
Sledding	卌 |||
Snowboarding	卌 卌 卌
Ice Skating	卌 |

Answer __________

5. The table shows the hours of TV that the students watch each day. How many students were surveyed in all?

TV Watched Each Day

Number of Hours	Tally
0	||
1	卌 ||
2	卌 |
3	卌 |||

Answer __________

EXTENDED-RESPONSE QUESTION

6. The students recorded the color of each vehicle in the parking lot.

blue, white, red, blue, gray
white, blue, black, black, blue
gray, black, blue, red, red
red, black, black, white, black

Part A Complete a frequency table showing the data. Make sure to write a title.

Color	Tally

Part B How many more blue vehicles than gray vehicles are in the parking lot?

Answer __________

31 Pictographs

Getting the Idea

A **pictograph** uses pictures to display and compare data. The **key** tells how many each symbol represents.

Here are the parts of a pictograph.
This graph shows the number of books that each person read.

title → **Books Read at the Read-A-Thon**

Pete	📘📘📘
Rosa	📘📘📘📘
Emily	📘📘📘📘📘 ← symbols
Tyrone	📘📘📘📘

Key: Each 📘 = 2 books ← key

EXAMPLE 1

Use the pictograph above. How many books did Pete read?

STRATEGY **Look in Pete's row. Use the symbols to decide how many books he read.**

STEP 1 Find Pete's row. Count the symbols.

There are 3 symbols.

STEP 2 Use the key to find how many each symbol represents.

Each symbol equals 2 books.

STEP 3 Multiply to find how many books Pete read.

3 symbols $\times$ 2 books $=$ 6 books

SOLUTION **Pete read 6 books.**

EXAMPLE 2

Use the pictograph on page 212.

Who read the greatest number of books? How many books did that student read?

STRATEGY **Find the row with the greatest number of symbols. Use the key to find how many books the symbols represent.**

STEP 1 Find the row with most symbols.

The row for Emily shows 5 symbols.

Emily read the greatest number of books.

STEP 2 Find the number of books Emily read.

Each symbol stands for 2 books.

STEP 3 Use repeated addition or multiplication.

$2 + 2 + 2 + 2 + 2 = 10$ books

5 symbols $\times$ 2 books $=$ 10 books

SOLUTION **Emily read the greatest number of books.**
She read 10 books.

COACHED EXAMPLE

Monica asked her classmates which time of day they prefer. She recorded the results in the table below.

Time of Day	Number of Students
Morning	8
Afternoon	12
Evening	4

Complete the pictograph to represent Monica's data.

THINKING IT THROUGH

What would be a good title for the graph? ______________________

Label the title on the pictograph.

Choose a picture for the symbol. ____________

Decide on a number to use for the key.

Each symbol equals _____ students.

Label the key on the pictograph.

How many symbols are needed for the Morning row? _____

How many symbols are needed for the Afternoon row? _____

How many symbols are needed for the Evening row? _____

Draw the symbols on the pictograph.

Complete the graph.

____

Key: ______________________________

Lesson Practice

Choose the correct answer.

Use the pictograph for questions 1–3.

Bags of Cookies Sold

Chocolate Chip	●●●●●
Oatmeal Raisin	●●
Peanut Butter	●●●
Mint	●●●●

Key: Each ● = 5 bags of cookies

1. What does each symbol in the pictograph represent?

A. 5 jars

B. 5 scoops

C. 5 cups

D. 5 bags

2. How many bags of mint cookies were sold?

A. 20 **C.** 9

B. 16 **D.** 4

3. How many more bags of chocolate chip cookies were sold than bags of oatmeal raisin cookies?

A. 3 **C.** 15

B. 6 **D.** 35

Use the pictograph for questions 4–6.

Students' Favorite Music

Pop	♫♫♫♫♫
Country	♫♫♫♫♫♫♫
Blues	♫♫♫♫
Classical	♫♫

Key: Each ♫ = 2 students

4. How many students chose country as their favorite type of music?

A. 4 **C.** 10

B. 8 **D.** 14

5. Which type of music did 8 students choose?

A. pop

B. country

C. blues

D. classical

6. What is the total number of students that chose a favorite type of music?

A. 38

B. 36

C. 34

D. 18

7. How many more tigers are there than elephants?

Animals at a Zoo

Bears	🐻🐻🐻🐻🐻🐻
Elephants	🐻🐻½
Giraffes	🐻🐻🐻🐻🐻
Tigers	🐻🐻🐻½

Key: Each 🐻 = 2 animals

Answer ________

EXTENDED-RESPONSE QUESTION

8. Vera spends time practicing the piano. The table shows the amount of time she spent practicing in 4 days.

Time Spent Practicing Piano

Day	Monday	Wednesday	Thursday	Saturday
Time Spent Practicing	10 min	20 min	15 min	30 min

Part A Complete the pictograph below to show the data in the table.

____

Key: ________________

Part B How many minutes did Vera spend in all practicing the piano?

Answer ________ minutes

Lesson

32 Bar Graphs

3.S.4, 3.S.5, 3.S.6, 3.S.7, 3.S.8

Getting the Idea

A **bar graph** uses bars of different lengths to compare data. The **scale** tells how many each bar represents.

To read the value of a bar, find the line that lines up with the top of the bar. Follow that line to the scale to read the number. If the bar ends between two lines, estimate the value based on the two numbers it ends between.

Here are the parts of a bar graph.

This graph shows the number of students absent each day this week.

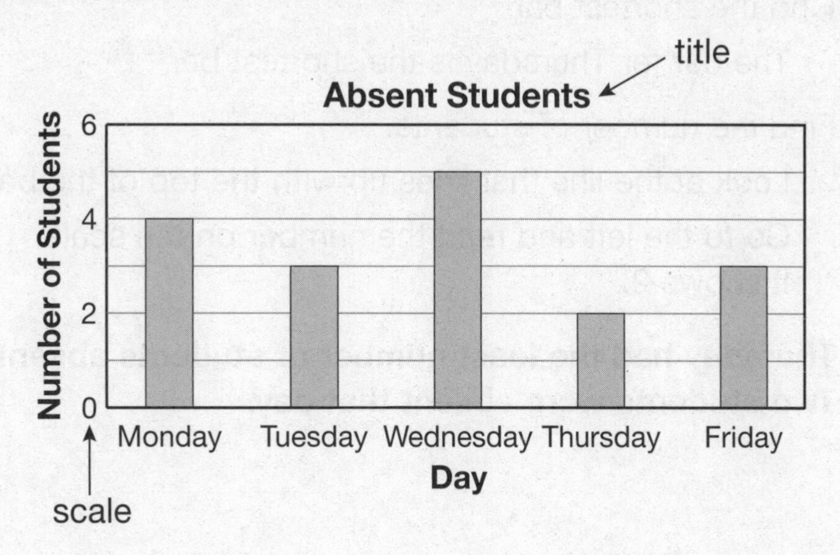

EXAMPLE 1

On which day was the least number of students absent?
How many students were absent that day?

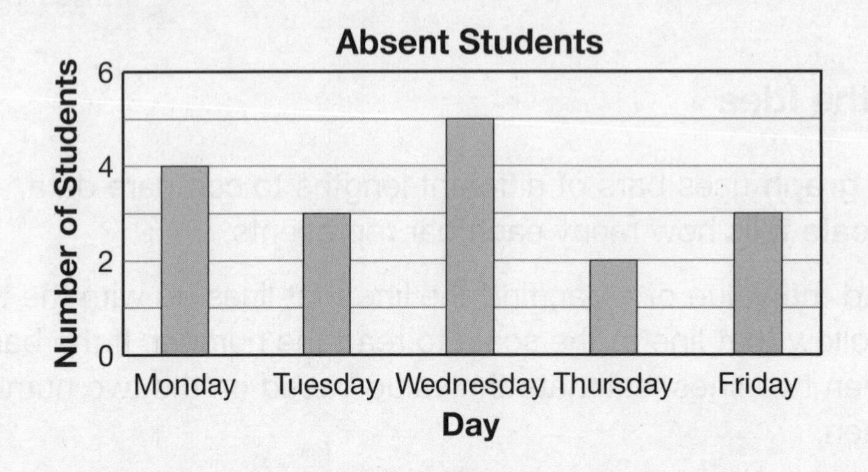

STRATEGY **Find the shortest bar. Then look at the scale.**

STEP 1 Find the shortest bar.

The bar for Thursday is the shortest bar.

STEP 2 Find the number of students.

Look at the line that lines up with the top of the bar.

Go to the left and read the number on the scale.
It shows 2.

SOLUTION **Thursday had the least number of students absent.**
Two students were absent that day.

EXAMPLE 2

The bar graph shows the number of members in four different clubs. How many more members are in the band than in the math club?

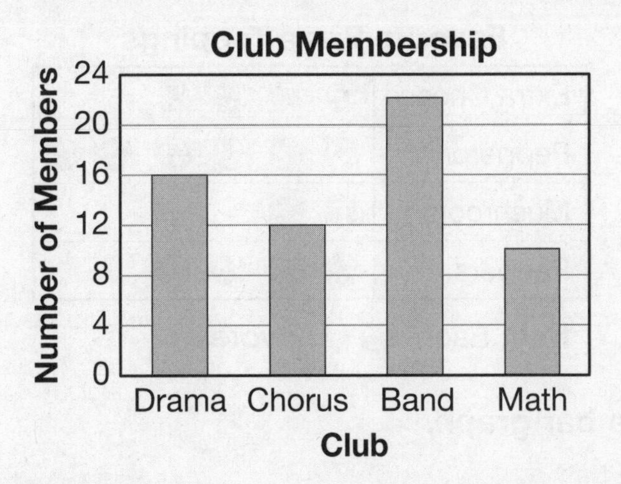

STRATEGY **Find the number of members in each club.**

STEP 1 Find the number of band members.

The bar ends halfway between 20 and 24.

Halfway between 20 and 24 is 22.

There are 22 band members.

STEP 2 Find the number of math club members.

The bar ends halfway between 8 and 12.

Halfway between 8 and 12 is 10.

There are 10 math club members.

STEP 3 Subtract.

$22 - 10 = 12$

SOLUTION **There are 12 more members in the band than in the math club.**

EXAMPLE 3

Isaac asked his classmates which pizza topping they prefer. The pictograph shows the data. Make a bar graph showing the same data as the pictograph.

Favorite Pizza Toppings

Topping	Pizzas
Extra Cheese	🍕 🍕 🍕 🍕
Pepperoni	🍕 🍕 🍕 🍕 🍕 🍕
Mushrooms	🍕 🍕
Peppers	🍕 🍕 🍕 🍕

Key: Each 🍕 = 2 votes

STRATEGY **Make a bar graph.**

STEP 1 Label the graph.

Write the title: Favorite Pizza Toppings.

Label the side of the graph. Use a scale of 2.

Label the bottom side with the names of the toppings.

STEP 2 Find the numbers of votes for each topping.

Extra cheese has 4 pizzas. It has 8 votes.

Pepperoni has 6 pizzas. It has 12 votes.

Mushrooms has 2 pizzas. It has 4 votes.

Peppers has 4 pizzas. It has 8 votes.

STEP 3 Draw the bars.

SOLUTION **The bar graph shows the same data as the pictograph.**

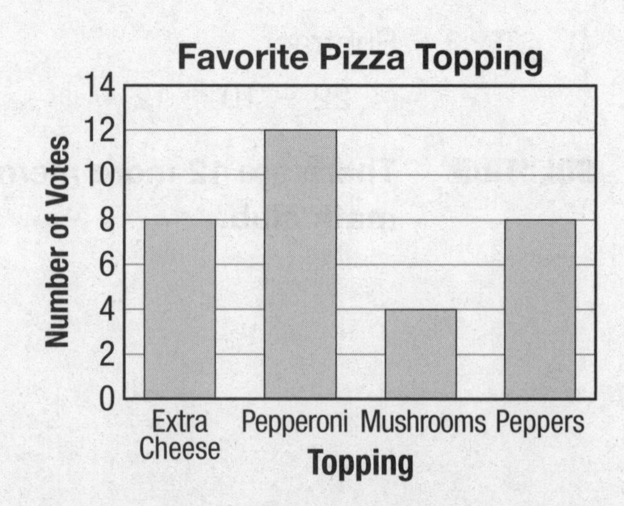

COACHED EXAMPLE

Students in the third grade voted on a school nickname. One student's vote did not get on the graph.

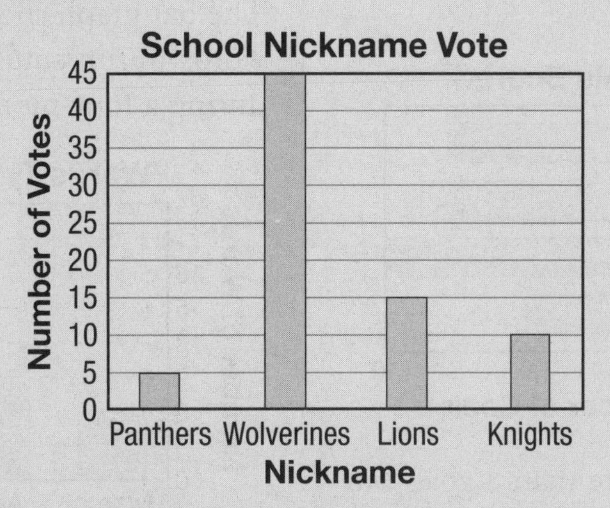

Based on the data in the graph, which nickname did this student most likely vote for?

THINKING IT THROUGH

Look at the bars in the graph.

Find the number of votes for each nickname.

Panthers received _____ votes.

Wolverines received _____ votes.

Lions received _____ votes.

Knights received _____ votes.

The nickname that received the most votes will most likely be the nickname the student voted for.

Which nickname received the most votes? ______________________________

Based on the data in the graph, the student most likely voted for _____________________.

Lesson Practice

Choose the correct answer.

Use the bar graph for questions 1–3.

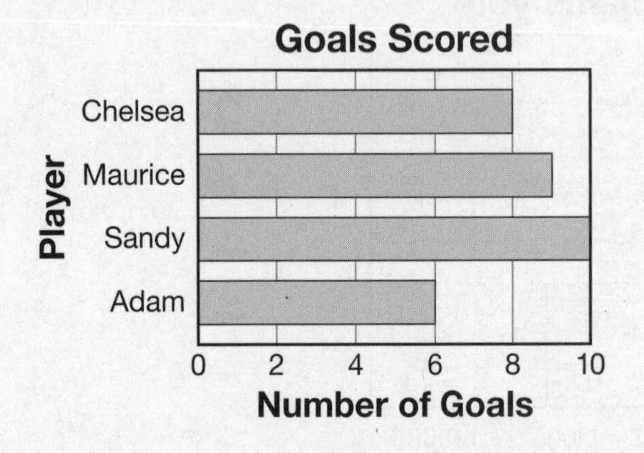

1. Who scored more than 8 goals but fewer than 10 goals?

A. Adam

B. Sandy

C. Maurice

D. Chelsea

2. Who scored 8 goals?

A. Adam

B. Sandy

C. Maurice

D. Chelsea

3. How many goals did Maurice and Adam score in all?

A. 14 **C.** 16

B. 15 **D.** 18

Use the bar graph for questions 4 and 5.

The bar graph shows the number of words Karim could type in one minute during a four-month period.

4. The data in the graph follow a pattern. What is the rule for the pattern?

Answer ______________________

____

5. Based on the data in the graph, how many words will Karim be able to type in one minute in July?

Answer ____________

EXTENDED-RESPONSE QUESTION

6. The table shows the students' votes on their favorite fruits.

Favorite Fruits

Fruit	Number of Students
Banana	6
Orange	9
Apple	1
Grape	8

Part A Complete the bar graph below to show the data in the table.

Part B How many more students chose oranges as their favorite fruit than chose bananas?

Answer ____________

Lesson

33 Collect and Record Data

3.S.1, 3.S.2

Getting the Idea

A **survey** is a way of collecting data by asking people questions. You can use a table to collect the data and use a graph to display it.

EXAMPLE 1

The table shows the results of Hanna's survey. Which question did Hanna most likely ask?

Hanna's Survey

Sport	Tally
Baseball	𝍸 𝍸
Football	𝍸 𝍸 |||
Basketball	𝍸 𝍸 𝍸 |||
Soccer	𝍸 ||

A. Do you like to watch sports?

B. What is your favorite type of movie?

C. What time does the game start?

D. Which is your favorite sport to watch?

STRATEGY **Review the question in each choice.**

STEP 1 Choice A: This would have the answers of yes and no.

STEP 2 Choice B: The data in the table is not about movies.

STEP 3 Choice C: The data in the table is not about time.

STEP 4 Choice D: The data in the table is about types of sports.
This question makes sense.

SOLUTION **Hanna most likely asked the question in Choice D, "Which is your favorite sport to watch?"**

EXAMPLE 2

Make a bar graph using the data in the table in Example 1.

STRATEGY **Use the data in the table.**

STEP 1 Find the number of people for each sport.

Add an extra column. Title it "Number of People."

Sport	Tally	Number of People
Baseball	𝍸 𝍸	10
Football	𝍸 𝍸 |||	13
Basketball	𝍸 𝍸 𝍸 |||	18
Soccer	𝍸 ||	7

STEP 2 Make a graph.

Write a title. The title can be "Favorite Sport to Watch."

Label the side of the graph. Use a scale of 2.

Label the bottom of the graph. Write the names of the sports.

Draw the bars to the length shown by the number in the table.

SOLUTION

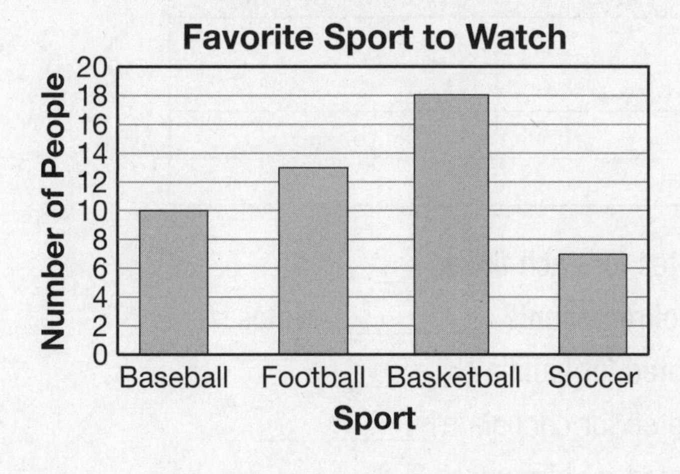

COACHED EXAMPLE

Scarlett did a survey. She made a pictograph to show the data.

Favorite Ice Cream Flavors

Vanilla	🍦🍦🍦🍦🍦🍦🍦
Chocolate	🍦🍦🍦🍦🍦🍦
Strawberry	🍦🍦🍦🍦
Mint	🍦🍦🍦

Key: Each 🍦 = 4 votes

Make a table to show the same data as in the pictograph.
Write a question that Scarlett most likely asked for the survey.

THINKING IT THROUGH

Make a table.

Write the title and names of flavors.

____

Flavor	Number of Votes

Find the number of votes for each flavor.

What does each symbol represent? __________ votes

How many students voted for vanilla? __________

How many students voted for chocolate? __________

How many students voted for strawberry? __________

How many students voted for mint? __________

Fill the numbers in the table.

Write a question that Scarlett most likely asked.

____

Lesson Practice

Choose the correct answer.

Use the table for questions 1–3.

Jin Lee did a survey and made the table below.

Favorite Lunch Food

Food	Number of Students
Hamburger	12
Pizza	20
Hot Dogs	12
Grilled Cheese	8

1. Which question did Jin Lee most likely ask?

   A. How much did you spend on lunch?

   B. Do you like to have pizza for lunch?

   C. What is your favorite lunch food?

   D. At what time do you eat lunch?

2. How many people did Jin Lee survey about favorite lunch food?

   A. 20

   B. 32

   C. 50

   D. 52

3. Which pictograph shows the same data as the table?

A.

Favorite Lunch Food

Hamburger	○○○○○○
Pizza	○○○○○○○○○○
Hot Dogs	○○○○○○
Grilled Cheese	○○○○

Key: Each ○ = 4 students

B.

Favorite Lunch Food

Hamburger	○○
Pizza	○○○
Hot Dogs	○○○○○
Grilled Cheese	○○○

Key: Each ○ = 4 students

C.

Favorite Lunch Food

Hamburger	○○○
Pizza	○○○○○
Hot Dogs	○○○
Grilled Cheese	○○

Key: Each ○ = 4 students

D.

Favorite Lunch Food

Hamburger	○○○
Pizza	○○○
Hot Dogs	○○
Grilled Cheese	○○○○○

Key: Each ○ = 4 students

Use the bar graph for questions 4 and 5.

Julius recorded the number of students who volunteered at the park each day. He made the bar graph below.

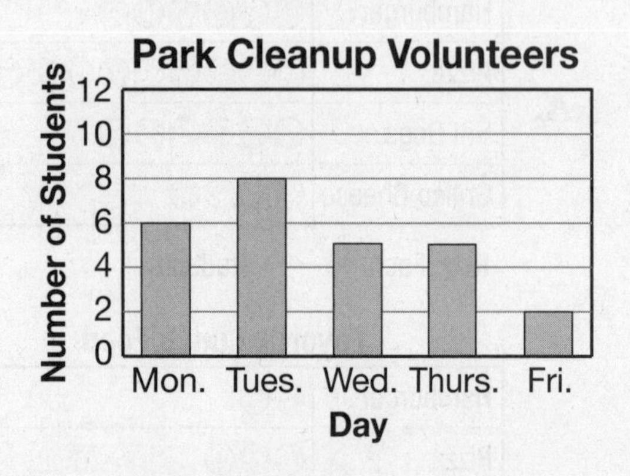

4. How many more volunteers were there on Monday and Tuesday than on Thursday and Friday?

Answer ________

5. How many students in all volunteered for the park cleanup in those 5 days?

Answer ________

EXTENDED-RESPONSE QUESTION

6. Mrs. Thomas surveyed some students and made a graph.

Favorite Type of Bread

White	BREAD BREAD BREAD BREAD BREAD
Wheat	BREAD BREAD BREAD BREAD
Rye	BREAD BREAD

Key: Each BREAD = 2 students

Part A Write a question that Mrs. Thomas most likely asked.

____

Part B Complete the table below to show the results in the pictograph. Be sure to write a title.

Type of Bread	Number of Students

5 Review

1 Stacy asked 24 families about the number of televisions they own.

The list below shows the number of TVs each family owns.

3	0	3	1	1	3
2	2	2	2	3	2
2	0	3	2	3	2
3	1	2	2	1	2

Which table shows the same data?

A

NUMBER OF TVs OWNED

Number of TVs	Tally
0	|
1	𝍸
2	𝍸 𝍸
3	𝍸 ||

B

NUMBER OF TVs OWNED

Number of TVs	Tally
0	||
1	𝍸
2	𝍸 𝍸 |
3	𝍸 |||

C

NUMBER OF TVs OWNED

Number of TVs	Tally
0	|
1	||||
2	𝍸 𝍸 ||
3	𝍸 |

D

NUMBER OF TVs OWNED

Number of TVs	Tally
0	||
1	||||
2	𝍸 𝍸 |
3	𝍸 ||

Use the pictograph for questions 2–4.

The pictograph shows the number of music CDs sold in a store.

CDs SOLD

Type of Music	Number of CDs
Rock	◎◎
Blues	◎◎◎
Jazz	◎◎
Pop	◎◎◎◎◎◎

KEY
◎ = 10 CDs

Key: Each ◎ = 10 CDs

2 How many jazz CDs were sold?

A 20

B 30

C 35

D 40

3 How many more pop CDs were sold than blues CDs?

A 3

B 30

C 40

D 60

4 Based on the data, which is **most likely** the next type of CD sold?

A rock

B blues

C jazz

D pop

Use the bar graph for questions 5–7.

Eliot did a survey and made this bar graph.

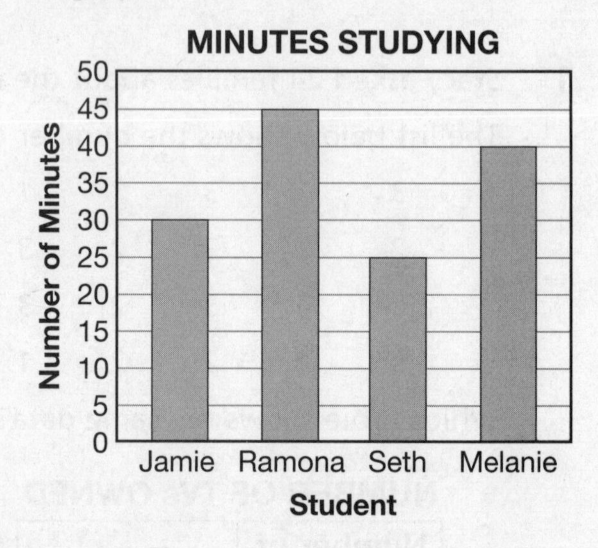

5 Which question did Eliot **most likely** ask?

A At what time did you study?

B What is your favorite time of day?

C What are you studying?

D How long did you spend studying?

6 Which student studied for 40 minutes?

A Jamie

B Ramona

C Seth

D Melanie

7 How many more minutes did Ramona study than Seth?

A 25

B 20

C 30

D 40

Use the bar graph for questions 8–10.

Students voted on the place they would like to go for the class trip.

The bar graph shows the results.

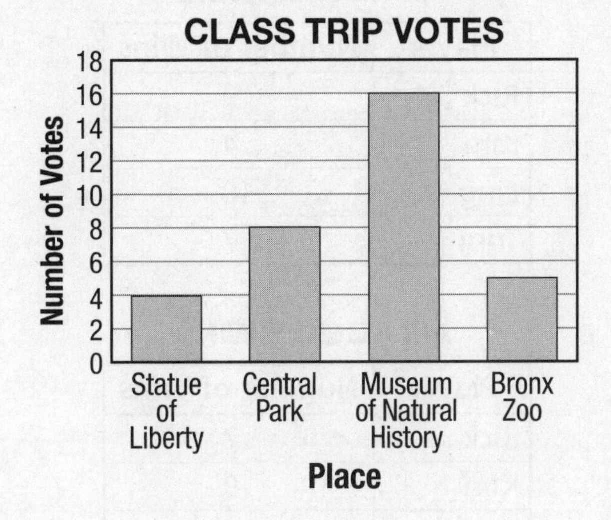

8 Which place received 5 votes?

A Bronx Zoo

B Museum of Natural History

C Central Park

D Statue of Liberty

9 How many votes did the least popular place receive?

A 2

B 4

C 5

D 8

10 The teacher will use the data to pick the class trip. Based on the data, which place will the teacher **most likely** pick for the class trip?

A Bronx Zoo

B Museum of Natural History

C Central Park

D Statue of Liberty

11 Tina recorded the air hockey wins of 4 players. She made the graph below.

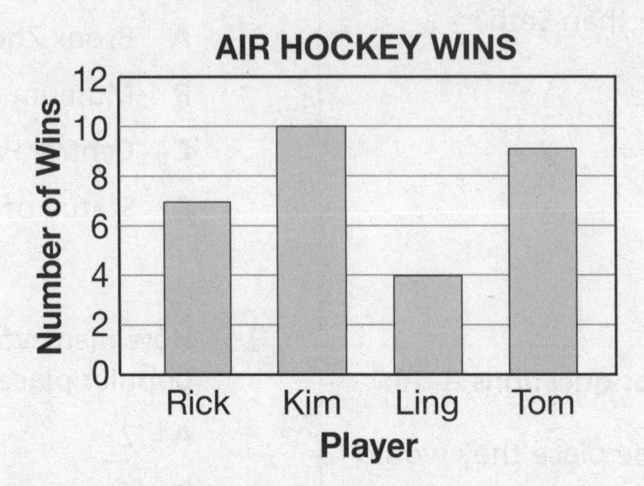

Which shows the same data as the bar graph?

A **AIR HOCKEY WINS**

Player	Number of Wins
Rick	7
Kim	10
Ling	4
Tom	9

B **AIR HOCKEY WINS**

Player	Number of Wins
Rick	6
Kim	10
Ling	5
Tom	8

C **AIR HOCKEY WINS**

Player	Number of Wins
Rick	9
Kim	4
Ling	10
Tom	7

D **AIR HOCKEY WINS**

Player	Number of Wins
Rick	7
Kim	9
Ling	4
Tom	10

12 The pictograph shows Joyner's hits in the fall softball season.

JOYNER'S HITS IN FALL SOFTBALL SEASON

Type of Hit	Number of Hits
Single	⚾ ⚾ ⚾ ⚾ ⚾
Double	⚾ ⚾
Triple	⚾ ⚾ ⚾
Homerun	⚾

KEY
⚾ = 3 hits

How many doubles did Joyner hit?

Answer ________________ doubles

13 The bar graph shows the books read by each grade in a week.

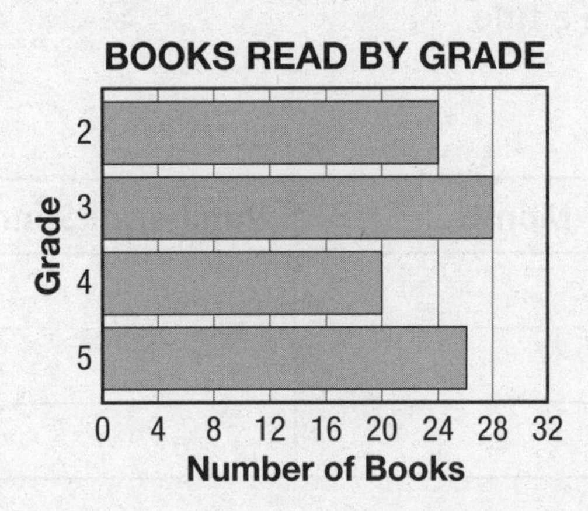

Which grade read more than 24 books, but fewer than 28 books?

Answer ________________

14 Velma kept track of the number of sunny days last spring. She made a pictograph showing the data.

SUNNY DAYS LAST SPRING

Month	Number of Days
April	☀ ☀ ☀ ☀ ☀ ☀ ☀
May	☀ ☀ ☀ ☀ ☀ ☀ ☀ ☀ ☀
June	☀ ☀ ☀ ☀ ☀ ☀ ☀ ☀ ☀ ☀

KEY
☀ = 2 sunny days

Part A

Complete the table below to show the number of sunny days in each month. Make sure to write a title.

Month	Number of Sunny Days

Part B

How many more sunny days were there in June than in April?

Answer ________________ days

Glossary

A

add (addition) an operation that combines two or more quantities to find the sum or total (Lesson 4)

addend the numbers being added in an addition problem (Lesson 4)

angle a figure formed where two lines meet (Lesson 21)

array shows equal groups in rows and columns (Lesson 7)

associative property of addition a property that states that the grouping of addends does not affect the sum (Lesson 9)

B

bar graph a graph that uses horizontal or vertical bars to display and compare data (Lesson 32)

C

capacity a measure that tells how much a container can hold (Lesson 27)

circle a two-dimensional shape with a curved side and 0 angles (Lesson 21)

commutative property of addition a property that states the order of the addends does not affect the sum (Lesson 9)

commutative property of multiplication a property that states that the order of the factors does not affect the product (Lesson 9)

cone a three-dimensional shape with a circular face (Lesson 23)

congruent figures figures that have the same size and same shape (Lesson 22)

cube a three-dimensional shape with 6 square faces (Lesson 23)

cup (c) a customary unit of capacity; 2 cups = 1 pint (Lesson 27)

cylinder a three-dimensional shape with a curved surface and 2 faces that are circles (Lesson 23)

D

data information (Lesson 30)

denominator the bottom number of a fraction; tells how many equal parts in a whole or in a group (Lesson 13)

difference the answer in a subtraction problem (Lesson 4)

digit 0, 1, 2, 3, 4, 5, 6, 7, 8, and 9 (Lesson 2)

dime a coin worth 10 cents (Lesson 28)

divide (division) operation on two numbers that tells how many groups or how many in each group **(Lesson 11)**

dividend a number to be divided by another number **(Lesson 11)**

divisor the number by which the dividend is divided **(Lesson 11)**

E

edge a line where two faces of a three-dimensional shape meet **(Lesson 23)**

equivalent fractions fractions that have different numerators and denominators that name the same value **(Lesson 15)**

estimate a number close to the exact answer **(Lesson 6)**

even number a number with 0, 2, 4, 6, or 8 in the ones place **(Lesson 5)**

expanded form a way of writing a number that shows the value of each digit **(Lesson 2)**

F

face one side of a three-dimensional shape **(Lesson 23)**

fact family a set of related facts that use the same numbers **(Lesson 11)**

factors the numbers multiplied in a multiplication sentence **(Lesson 7)**

foot (ft) a customary unit of length; 1 foot = 12 inches **(Lesson 25)**

fraction a number that names a part of a whole or a part of a group **(Lesson 13)**

frequency table a table that shows how often each item, number, or range of numbers occurs in a group of data **(Lesson 30)**

G

gallon (gal) a customary unit of capacity; 1 gallon = 4 quarts **(Lesson 27)**

geometric pattern a pattern with shapes **(Lesson 20)**

H

hexagon a polygon with 6 sides and 6 angles **(Lesson 21)**

hour a unit of time; 1 hour = 60 minutes **(Lesson 29)**

I

identity property of multiplication a property that states that when a number is multiplied by 1, the product is that number **(Lesson 9)**

inch (in.) a customary unit of length; 12 inches = 1 foot **(Lesson 25)**

inverse operations operations that undo each other; addition is the inverse operation of subtraction; multiplication is the inverse operation of division (Lesson 11)

is equal to (=) symbol used to tell that one quantity is the same as another quantity (Lessons 3, 18)

is greater than (>) symbol used to tell that one quantity is more than another quantity (Lessons 3, 18)

is less than (<) symbol used to tell that one quantity is less than another quantity (Lessons 3, 18)

K

key a note in a pictograph that tells what each symbol in a graph represents (Lesson 31)

L

length the measure of how long, wide, or tall something is (Lesson 25)

line of symmetry a line that divides a figure into 2 equal halves (Lesson 24)

M

mile (mi) a customary unit of length; 1 mile = 1,760 yards (Lesson 25)

minuend the number that is being subtracted from in a subtraction problem (Lesson 4)

minute a unit of time; 60 minutes = 1 hour (Lesson 29)

multiply (multiplication) an operation that joins equal groups; a form of repeated addition (Lesson 7)

N

nickel a coin worth 5 cents (Lesson 28)

number pattern a pattern that uses only numbers (Lesson 19)

numerator the top number in a fraction (Lesson 13)

O

odd number a number with 1, 3, 5, 7, or 9 in the ones place (Lesson 5)

operation a process, such as addition, subtraction, multiplication, or division, performed on numbers (Lesson 17)

ounce (oz) a customary unit of weight; 16 ounces = 1 pound (Lesson 26)

P

pattern a set of numbers or shapes that uses a rule (Lesson 19)

penny a coin worth 1 cent (Lesson 28)

pictograph a graph that uses pictures or symbols to display data (Lesson 31)

pint (pt) a customary unit of capacity; 1 pint = 2 cups **(Lesson 27)**

place value the value indicated by the position of a digit in a number **(Lesson 2)**

place value chart a table that shows the value of each digit in a number **(Lesson 2)**

place value system a number system in which the position of a digit in a number determines its value **(Lesson 2)**

pound (lb) a customary unit of weight; 1 pound = 16 ounces **(Lesson 26)**

prediction an educated guess as to what will happen **(Lesson 30)**

product the answer in a multiplication problem **(Lesson 7)**

Q

quart (qt) a customary unit of capacity; 1 quart = 2 pints **(Lesson 27)**

quarter a coin worth 25 cents **(Lesson 28)**

quotient the answer in a division problem **(Lesson 11)**

R

rectangle a two-dimensional shape with 4 sides and 4 square corners **(Lesson 21)**

rectangular prism a prism with six faces that are triangles, 12 edges, and 8 vertices **(Lesson 23)**

regroup to rename a number in a different way **(Lesson 4)**

rhombus a parallelogram with equal sides **(Lesson 21)**

round estimate a number to the nearest ten or nearest hundred **(Lesson 6)**

rounding rules look to the digit to the right of the place you are rounding to. If the digit is less than 5, round down. If the digit is 5 or greater, round up. **(Lesson 6)**

rule tells how the numbers or symbols in a pattern are related **(Lesson 19)**

S

scale the part of the bar graph that gives the value of each bar **(Lesson 32)**

similar figures figures that have the same shape but not the same size **(Lesson 22)**

skip counting counting forward or backward by a number other than 1 **(Lesson 1)**

solid figure another term for a three-dimensional shape **(Lesson 23)**

sphere a three-dimensional shape with a curved surface and 0 faces (Lesson 23)

square a quadrilateral with 4 equal sides and 4 right angles (Lesson 21)

standard form a way of writing a number using digits (Lesson 2)

subtract (subtraction) an operation that takes a quantity away to find the difference (Lesson 4)

subtrahend the number that is taken away from another number in a subtraction problem (Lesson 4)

sum the answer in an addition problem (Lesson 4)

survey a way to gather information by asking questions (Lesson 33)

symmetry a shape has symmetry if it can be divided into two equal parts (Lesson 24)

T

three-dimensional shape a shape with length, width, and height; a solid figure (Lesson 17)

trapezoid a quadrilateral with exactly one pair of parallel sides (Lesson 21)

triangle a polygon with 3 sides and 3 angles (Lesson 21)

triangular prism a three-dimensional shape with 2 triangular faces and 3 rectangular faces (Lesson 23)

two-dimensional shape a shape with length and width (Lesson 21)

U

unit fraction a fraction with 1 in the numerator (Lesson 13)

V

vertex (plural: vertices) the point where 3 or more edges meet in a three-dimensional shape; the point at the top of a cone (Lesson 23)

W

weight the measure that tells how heavy something is (Lesson 26)

word form a way of showing a number using words (Lesson 2)

Y

yard (yd) a customary unit of length; 1 yard = 3 feet (Lesson 25)

Z

zero property of multiplication a property that states that when a number is multiplied by 0, the product is 0 (Lesson 9)

New York State Coach, Empire Edition, Mathematics, Grade 3

COMPREHENSIVE REVIEW 1

Name: ____________________

TIPS FOR TAKING THE TEST

Here are some suggestions to help you do your best:

- Be sure to read carefully all the directions in the test book.
- Read each question carefully and think about the answer before writing your response.
- Be sure to show your work when asked. You may receive partial credit if you have shown your work.

This picture means that you will use your ruler.

Session 1

1 A wedding had 206 guests. Which number is the same as 206?

A six hundred two

B two hundred sixty

C six hundred six

D two hundred six

2 Which number belongs on the line below to make the sentence true?

$$\_\_\_\_ < 359$$

A 360

B 501

C 359

D 347

3 Which number belongs in the blank to make the sentence true?

$$9 \times \_\_\_\_ = 9$$

A 0

B 1

C 9

D 10

4 There are 12 eggs in a dozen. Claude has 8 dozen eggs. How many eggs does he have?

A 80

B 84

C 86

D 96

5 Which shows $\frac{1}{3}$ of the rectangle white?

A

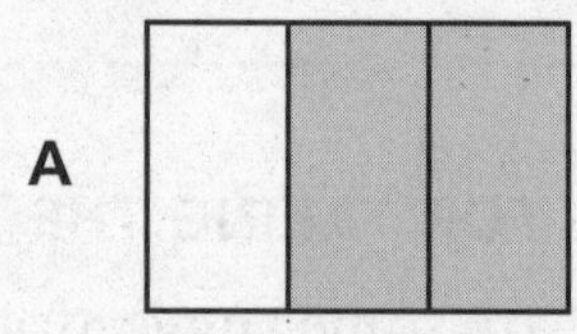

B

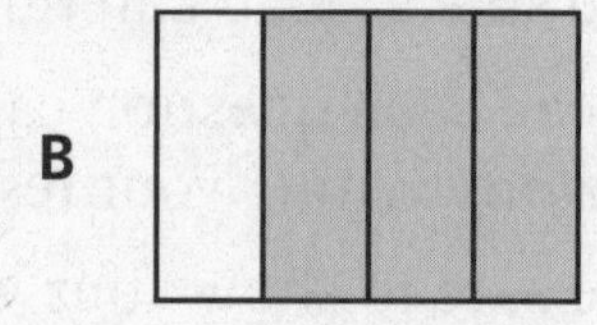

C

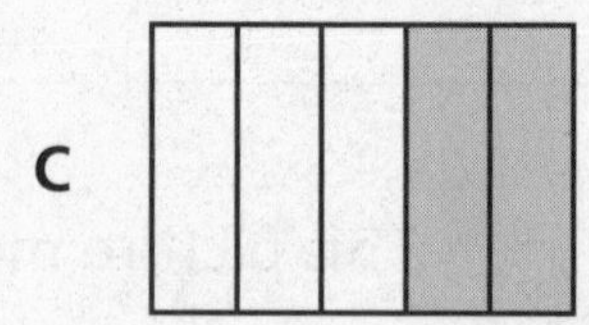

D

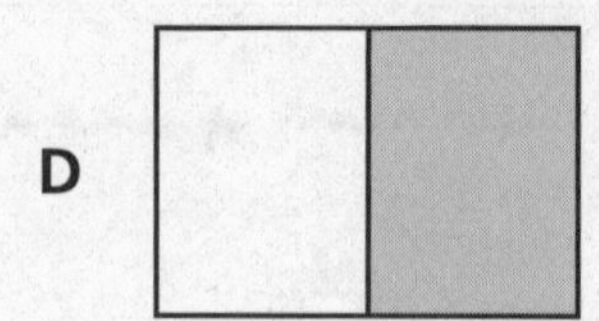

Go On

6 Lyle wrote this number sentence.

$12 \times 0 = \square$

What is the product of Lyle's number sentence?

A 0

B 1

C 12

D 120

7 Carina has 12 glasses to put on 3 shelves. She wants to put the same number of glasses on each shelf.

How many glasses will she put on each shelf?

A 4

B 5

C 8

D 15

8 Which **most likely** weighs about 1 pound?

A

B

C

D

Go On

9  Use your ruler to help you solve this problem.

How many inches long is this grasshopper?

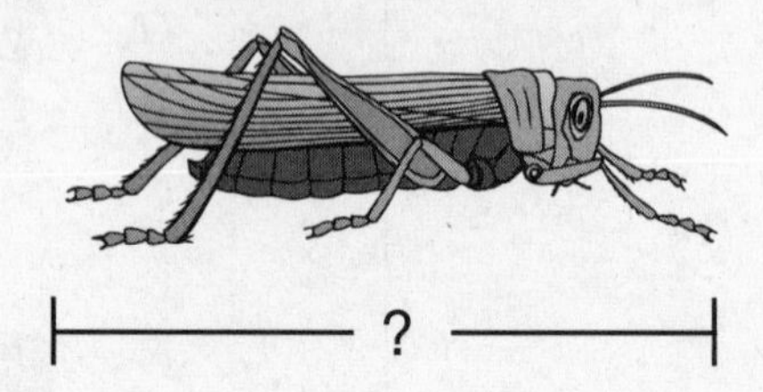

A 2 inches

B $2\frac{1}{2}$ inches

C 3 inches

D $3\frac{1}{2}$ inches

10 What is the next shape in this pattern?

A

B

C

D

Go On

11 Jared has these blocks in a bag.

What fraction of the blocks are marked with the letter A?

A $\frac{1}{10}$

B $\frac{1}{9}$

C $\frac{8}{9}$

D $\frac{1}{11}$

12 Blair wrote a number sentence.

$(6 + 3) + 4 = 3 + (\_\_\_\_ + 4)$

Which number makes the sentence true?

A 3

B 4

C 6

D 9

13 Terry went to the mall at 40 minutes past 4 o'clock. Which clock correctly shows the time that Terry went to the mall?

A 

B 

C 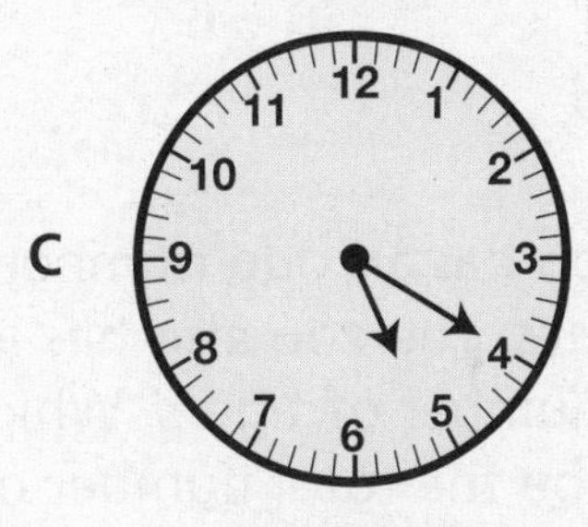

D 

Go On

14 The graph shows the types of cookies Marisa's Bakery made.

How many batches of chocolate chip cookies and peanut butter cookies did the bakery make?

A 18 **C** 8

B 14 **D** 6

15 The zoo has an odd number of monkeys. The zoo also has an even number of tigers. Which could be the total number of monkeys and tigers at the zoo?

A 20 **C** 37

B 34 **D** 42

16 Rochelle has 135 stickers in her collection. Erica has 98 stickers in her collection. How many more stickers does Rochelle have than Erica?

A 37

B 43

C 47

D 233

17 Don made this number pattern.

43, 51, 59, _?_, 75, 83

What is the missing number in Don's pattern?

A 61 **C** 67

B 63 **D** 71

18 Which shows the correct line of symmetry?

A

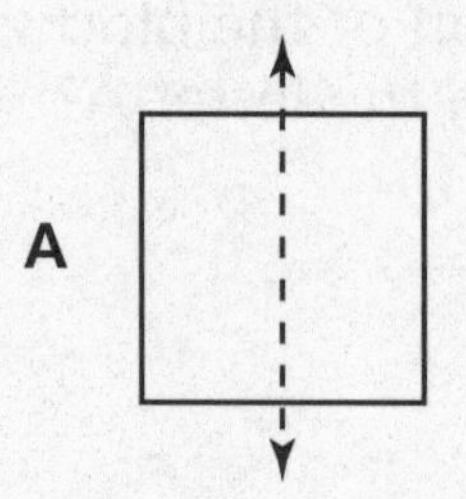

B

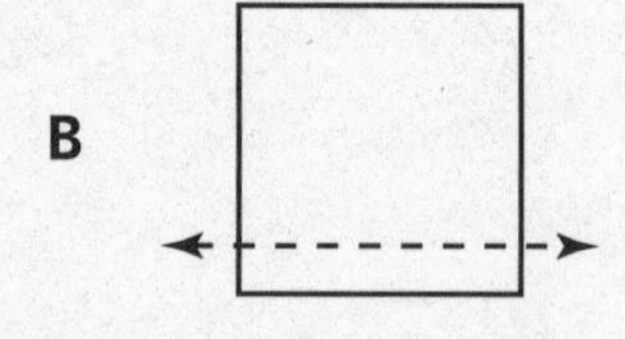

C

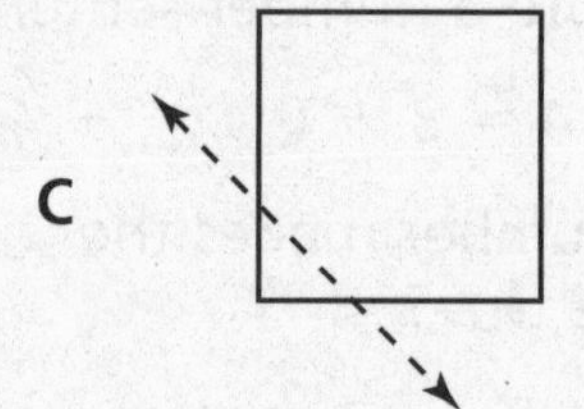

D

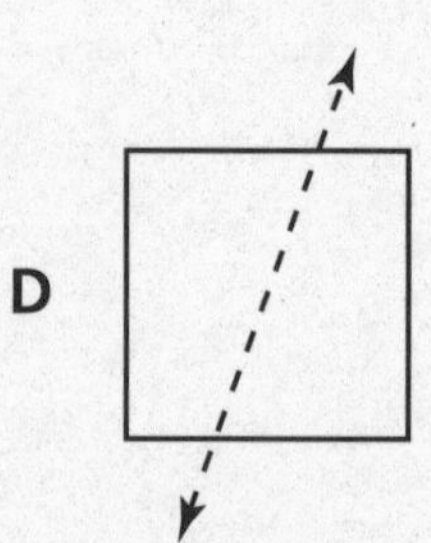

Go On

19 Mr. Robinson bought 6 tickets to the movies. Each ticket cost $8. How much did Mr. Robinson pay for the 6 tickets?

A $14

B $40

C $42

D $48

20 Which shows two congruent rectangles?

A
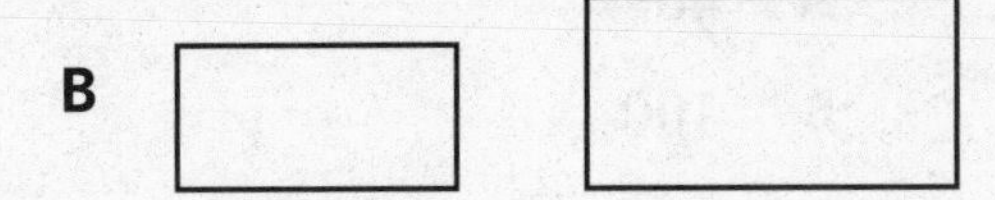

B
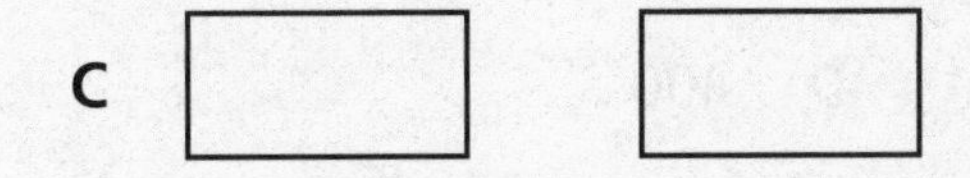

C
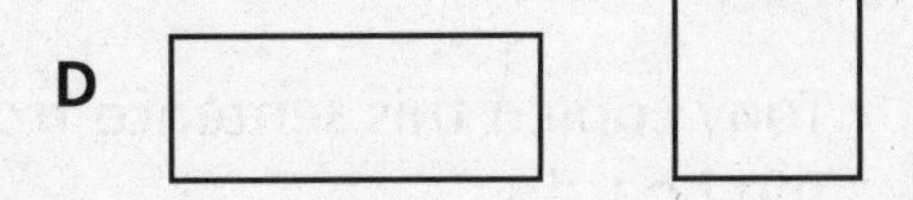

D

21 Kelly bought a small drink using the money shown.

How much did Kelly spend to buy the drink?

A $0.55

B $0.73

C $0.81

D $0.91

Go On

22 There are 100 paper clips in a box. How many paper clips are in 4 boxes?

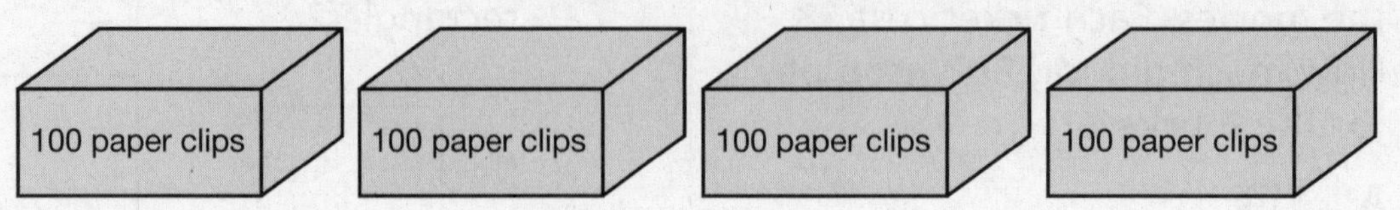

A 40

B 100

C 200

D 400

23 Tony copied this sentence from the board.

$\frac{1}{5} > \_\_\_\_$

Which fraction belongs on the line to make the sentence correct?

A $\frac{1}{3}$

B $\frac{1}{2}$

C $\frac{1}{5}$

D $\frac{1}{6}$

24 Shannon used a funnel to pour water into a bottle.

Which name best describes the shape of the funnel?

A cylinder

B cone

C cube

D sphere

Go On

25 The table shows some students' favorite fruits.

FAVORITE FRUIT

Fruit	Number of Students
Bananas	8
Apples	3
Oranges	3
Grapes	9

Which graph correctly shows the data in the table?

A

C

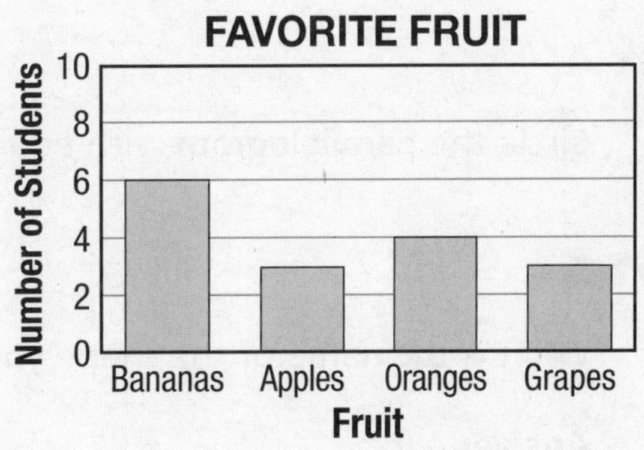

B

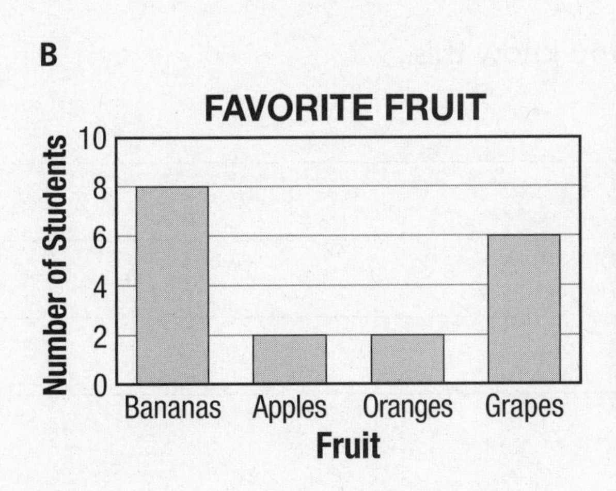

D

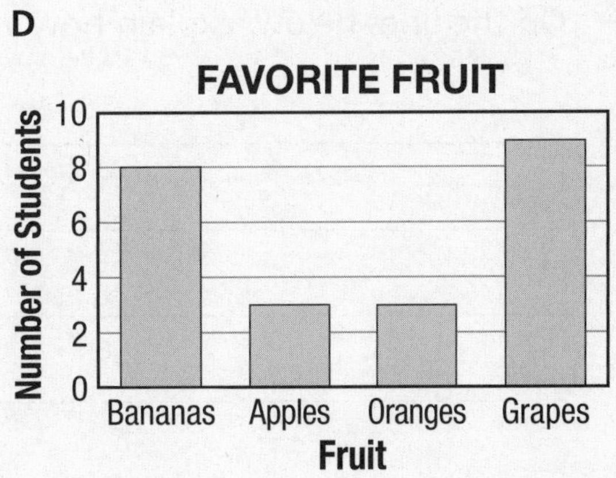

STOP

Session 2

26 Terri has the shapes below.

Part A

Circle the parallelogram with equal sides but no right angles.

Part B

What is the name of the shape you circled?

Answer ________________

On the lines below, explain how you know this.

____

____

____

____

Go On

27 Barry and Carla did a survey about how students get to school. The pictograph shows the numbers of students in each grade who ride bikes to school.

STUDENTS WHO BIKE TO SCHOOL

Grade	Number of Students
2nd Grade	🚲 🚲 🚲 🚲
3rd Grade	🚲 🚲 🚲
4th Grade	🚲 🚲 🚲 🚲 🚲 🚲
5th Grade	🚲 🚲 🚲 🚲

KEY
🚲 = 4 students

How many more 4th-grade students than 3rd-grade students ride their bikes to school?

Show your work.

Answer _______________ students

Go On

28 Tammy used 4 flowers for each bouquet she made. She made 3 bouquets.

How many flowers did Tammy use in all?

Show your work.

Answer ______________ flowers

29 Kayla received $250 for winning a writing contest. She bought a stereo for $159. She put the rest in the bank. How much did Kayla put in the bank?

Show your work.

Answer ______________

Go On

30 Evelyn worked at a drink stand one day. The table shows the drinks she sold.

Drink	Lemonade	Root Beer	Cola	Orange Soda
Number of Cups	?	12	12	8

Part A

Complete the pictograph to show the number of cups Evelyn sold for each drink. Be sure to use the key provided.

DRINKS SOLD

Drink	Number of Cups
Lemonade	🥤 🥤 🥤 🥤

KEY
🥤 = 4 cups

Part B

Based on the graph, how many more cups of lemonade than orange soda did Evelyn sell?

Answer _______________ cups

Go On

31 Melanie has collected 38 stamps from Spain and 53 stamps from France. Her goal is to collect a total of 200 stamps from both countries.

Part A

How many stamps has Melanie collected so far?

Answer _______________ stamps

Part B

About how many more stamps does Melanie need to reach her goal?

Answer _______________ stamps

On the lines below, explain the steps you took to find the answers.

____

____

____

STOP

New York State Coach, Empire Edition, Mathematics, Grade 3

COMPREHENSIVE REVIEW 2

Name: ______________________________

TIPS FOR TAKING THE TEST

Here are some suggestions to help you do your best:

- Be sure to read carefully all the directions in the test book.
- Read each question carefully and think about the answer before writing your response.
- Be sure to show your work when asked. You may receive partial credit if you have shown your work.

This picture means that you will use your ruler.

Session 1

1 A newspaper company made 382 copies of the Monday paper. They sold 682 copies of the Sunday paper. By how much is the number 682 greater than 382?

A 3 tens

B 3 hundreds

C 3 ones

D 3 thousands

2 Which number belongs on the line below to make the number sentence true?

$35 + 10 = \_\_\_\_\_ + 35$

A 45

B 35

C 20

D 10

3 Which container holds about 1 gallon?

A 

B 

C 

D 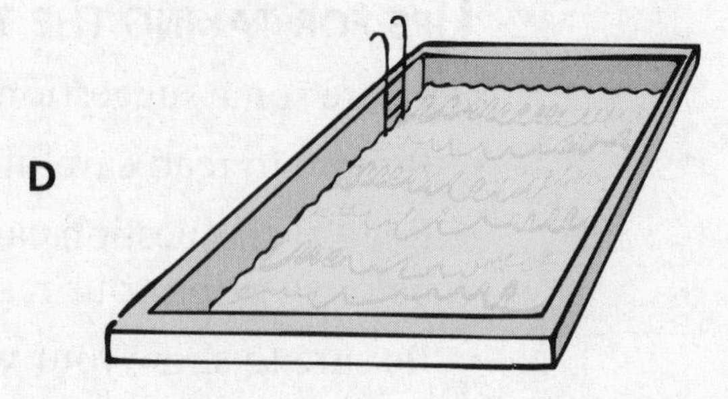

4 Which is the **best** estimate for the length of a DVD player remote control?

A 10 inches

B 10 feet

C 10 pounds

D 10 ounces

Go On

5 Grace made the pattern below.

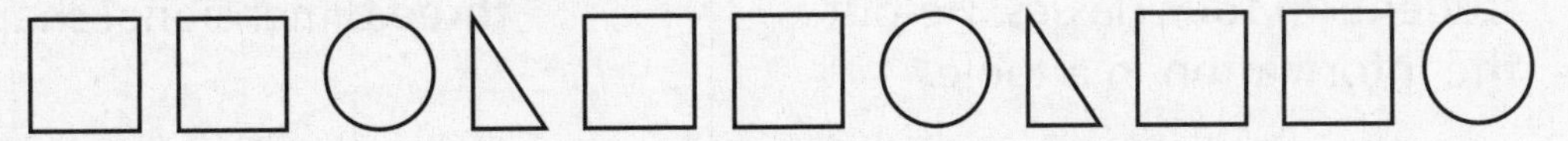

If the pattern continues, which shape should Grace draw next?

A □

B ○

C ◺

D ⏢

6 In which situation is an estimate as good as finding the exact answer?

A finding if you have enough money to buy two items

B finding if a chair is too wide to fit through a door

C finding what time you will get home from soccer practice

D finding how much money you will have left after buying an item

7 Harold picked oranges from 6 trees. He picked 5 oranges from each tree. How many oranges did Harold pick in all?

A 11

B 30

C 35

D 65

Go On

8 Nick counted the number of students in four classes. He put the information in a table.

THIRD-GRADE STUDENTS

Class	Number of Students
Math	28
Social Studies	21
English	27
P.E.	31

Which shows the best way to estimate the total number of students in social studies and English classes?

A 20 + 20 = □

B 20 + 30 = □

C 30 + 30 = □

D 30 − 20 = □

9 Jeff drew a face of a three-dimensional shape.

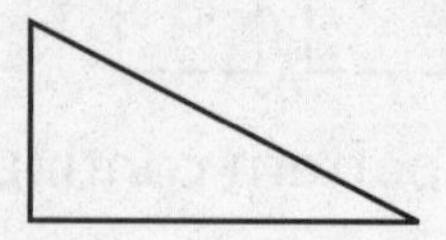

Which three-dimensional shape has this face?

A 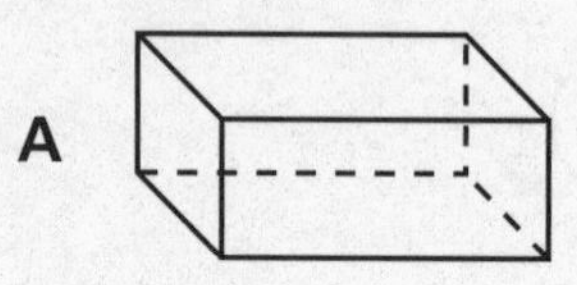

B 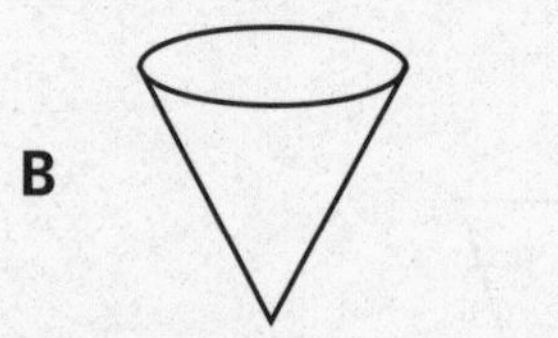

C 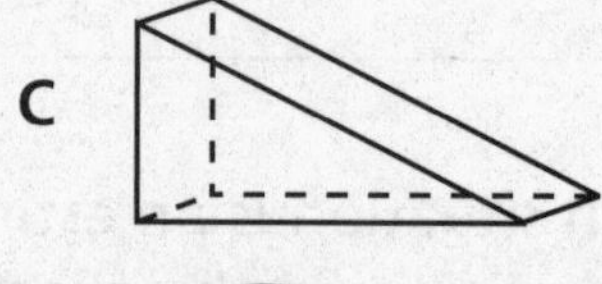

D 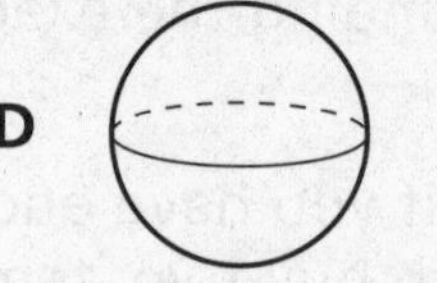

Go On

10 The bar graph shows the number of students in different clubs at school.

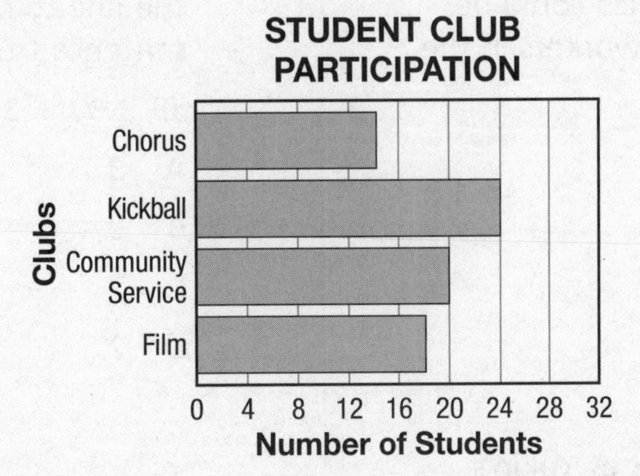

Which club has more than 16 but fewer than 20 members?

A Chorus

B Kickball

C Community Service

D Film

Go On

11 A toy store has an even number of workers. Which could be the number of workers in the toy store?

A 143

B 130

C 187

D 99

12 Different classes at Yukio's school volunteered at a local animal shelter. The table shows how much time each class spent helping at the animal shelter last year.

HOURS SPENT AT ANIMAL SHELTER

Class	Number of Hours
Grade 3	125
Grade 4	207
Grade 5	173
Grade 6	182

How many hours did Grade 4 and Grade 5 classes spend helping at the shelter last year?

A 280 hours

B 370 hours

C 380 hours

D 389 hours

13 What number belongs on the line to make the number sentence true?

$(9 + 7) + 3 = 9 + (\_\_\_\_ + 3)$

A 3

B 9

C 7

D 2

Go On

14 Which number belongs on the line to make the number sentence true?

$$\_\_\_ > 719$$

A 679

B 791

C 595

D 689

15 Which expression can be used to find the total number of counters below?

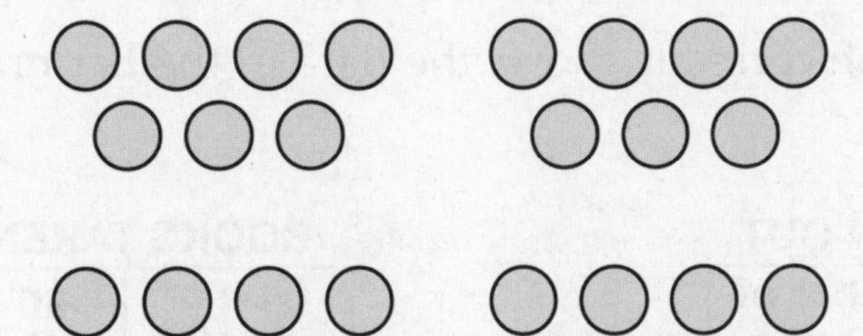

A 4×7

B $4 + 7$

C 7×7

D $6 + 7$

16 Kyle drew these figures in his notebook.

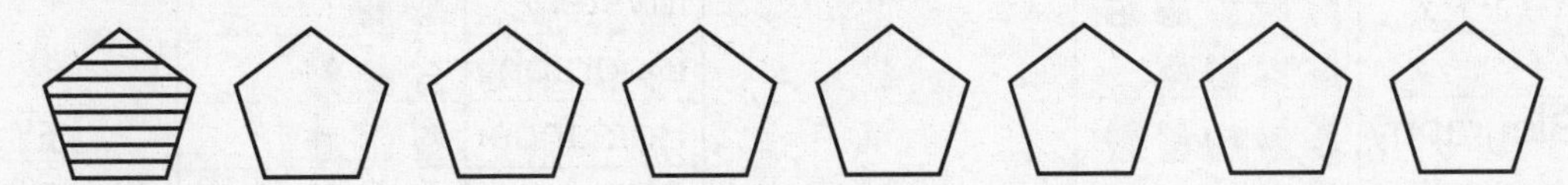

What fraction of the figures have stripes?

A $\frac{1}{8}$

B $\frac{1}{7}$

C $\frac{1}{6}$

D $\frac{7}{8}$

Go On

17 The bar graph shows books that were taken out of the library.

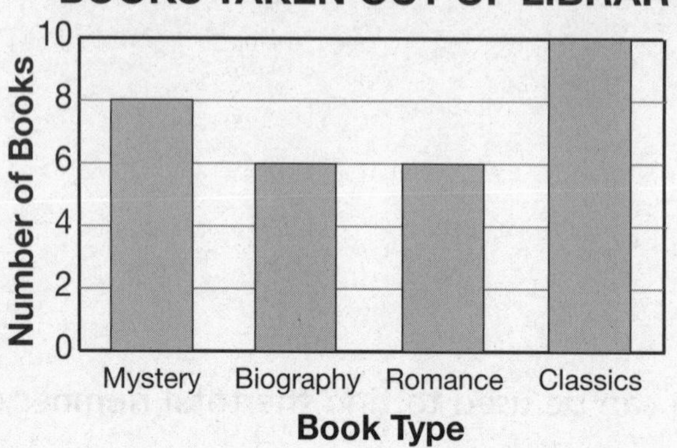

Which pictograph correctly shows the data in the bar graph?

A

BOOKS TAKEN OUT

Type of Book	Number of Books
Mystery	📕📕📕📕
Biography	📕📕📕
Romance	📕📕📕
Classics	📕📕📕📕📕

KEY
📕 = 2 books

B

BOOKS TAKEN OUT

Type of Book	Number of Books
Mystery	📕📕📕📕📕📕📕📕📕
Biography	📕📕📕📕📕📕
Romance	📕📕📕📕📕
Classics	📕📕📕📕📕📕📕📕📕📕

KEY
📕 = 2 books

C

BOOKS TAKEN OUT

Type of Book	Number of Books
Mystery	📕📕📕📕📕
Biography	📕📕
Romance	📕📕📕
Classics	📕📕📕

KEY
📕 = 2 books

D

BOOKS TAKEN OUT

Type of Book	Number of Books
Mystery	📕📕
Biography	📕📕📕
Romance	📕📕📕
Classics	📕📕📕📕📕

KEY
📕 = 2 books

Go On

18 Which number sentence is true?

A $472 < 427$

B $385 > 417$

C $105 = 55$

D $243 < 251$

19 Ms. Horne drew a pattern on the blackboard.

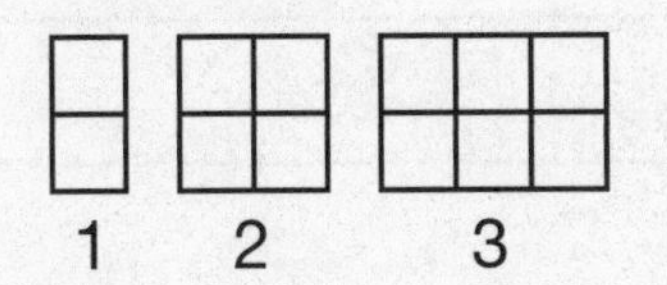

1 2 3 4

How many squares will be in the 4th figure?

A 4

B 8

C 10

D 12

20 Each bag contains 100 peanuts. How many peanuts are in 6 bags?

A 100

B 300

C 450

D 600

21 

Use your ruler to help you solve this problem.

The drawing shows a ribbon Yoli will use to tie a gift bag.

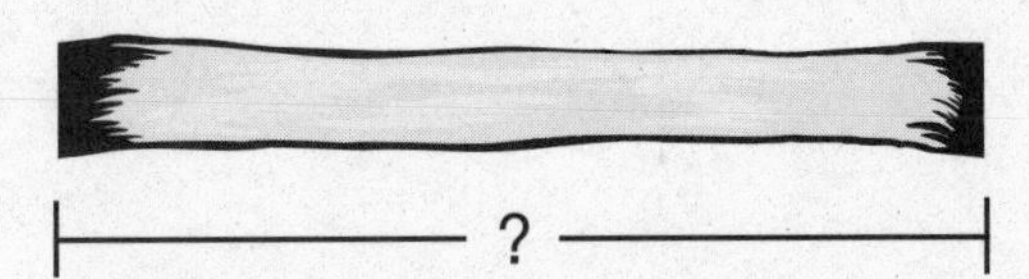

?

How many inches long is the ribbon?

A 2 inches

B $2\frac{1}{2}$ inches

C 3 inches

D $3\frac{1}{2}$ inches

22 Rick drew a diagonal line from the lower left to the upper right corner of a rectangle. What two polygons did he form?

A 2 triangles

B 2 rectangles

C 2 squares

D 2 trapezoids

Go On

23 Which shows the correct line of symmetry?

A 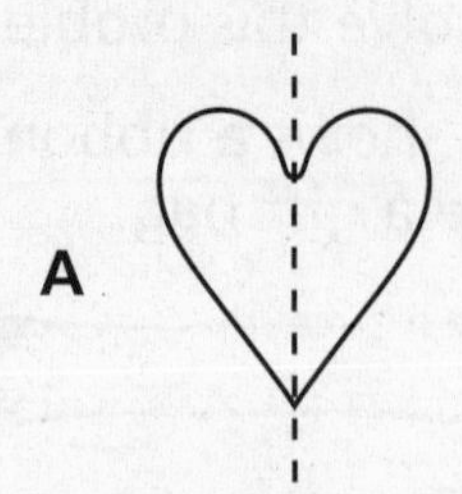

B 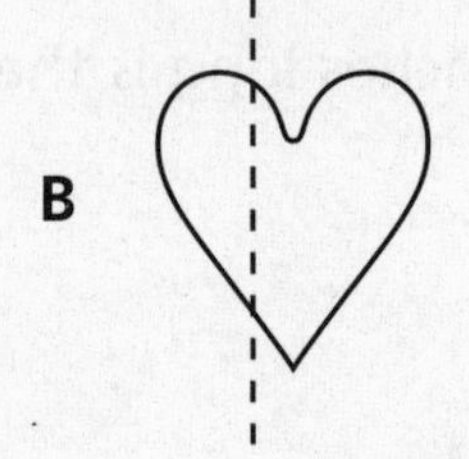

C 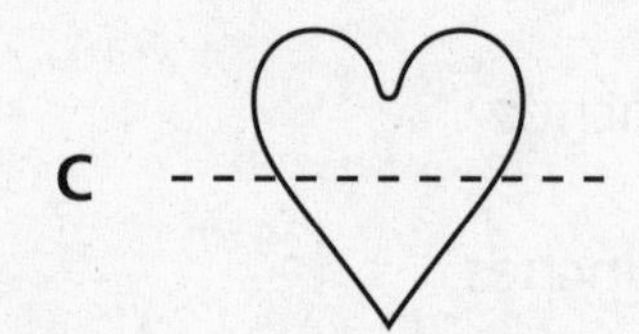

D 

24 Which shows $\frac{2}{5}$ of the rectangle shaded?

A

B

C

D

25 The clock shows the time that a movie starts. What time does the movie start?

A a quarter past four

B a quarter to four

C nine twenty

D a quarter past three

STOP

Session 2

26 Raven cut out stars for an art project. She cut out 15 stars. She put 5 stars on each piece of construction paper.

How many pieces of construction paper did Raven use?

Show your work.

Answer ______________ pieces

27 Melissa is reading a 437-page book. This week she read 171 pages. How many more pages does Melissa have left to read?

Show your work.

Answer ______________ pages

Go On

28 The pictograph shows the favorite horse colors of students in Ms. Carol's class.

FAVORITE HORSE COLORS

Color	Number of Students
Black	🐎🐎🐎
White	🐎🐎
Brown	🐎🐎🐎🐎🐎
Gray	🐎🐎

KEY
🐎 = 3 students

How many more students like brown horses than gray horses?

Show your work.

Answer ____________ students

Go On

29 Harumi drew the 3 shapes in the picture below.

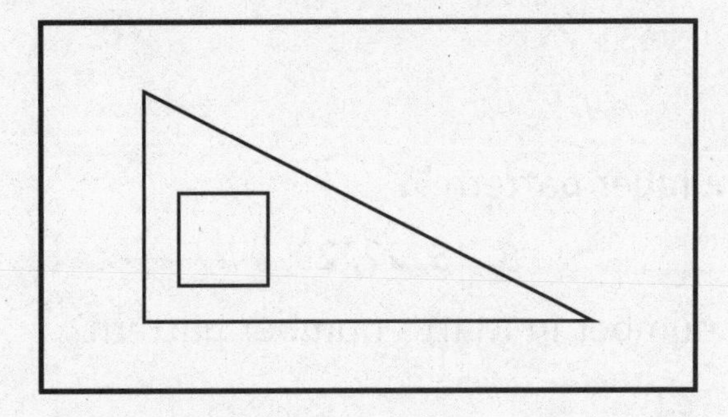

What is the name of the smallest shape?

Part A

Answer ________________

Part B

On the lines below, explain one way that all 3 shapes are alike.

____

____

____

Go On

30 Matthew made two types of patterns.

Part A

First he made a number pattern.

8, 15, 22, 29, ___?___

What is the next number in Matt's number pattern?

Answer ____________________

Part B

Then he drew a pattern with shapes.

⬡ ⬡ △ △ ⬡ ⬡ △ △ ⬡ ___?___ ___?___

On the lines below, draw the next two figures to continue Matt's pattern.

______________ ______________

On the lines below, explain how you know which two figures to draw.

____

____

____

Go On

31 Mr. Spinelli's class voted for their favorite pets. All the students wrote their favorite pets in the table below.

Dog	Cat
Cat	Bird
Fish	Dog
Dog	Cat
Dog	Dog
Fish	Fish
Fish	Dog
Dog	Cat

Part A

Complete the bar graph below to show the number of students who voted for each pet.

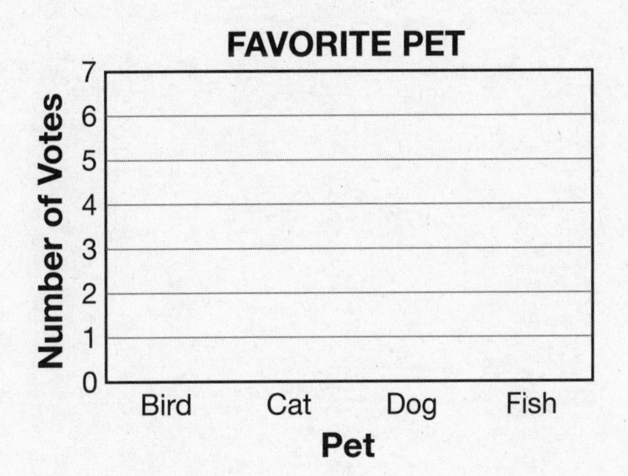

Part B

Based on the bar graph, how many students in all voted for their favorite pets?

Answer ______________ students

STOP

Punch-Out Tools

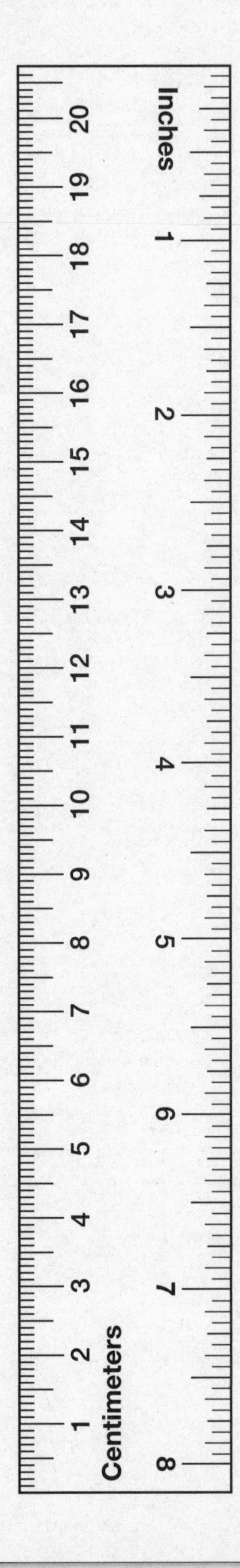

Notes